Web轻量级框架 Spring+Spring MVC+MyBatis整合开发实战

（第2版）

黄文毅 编著

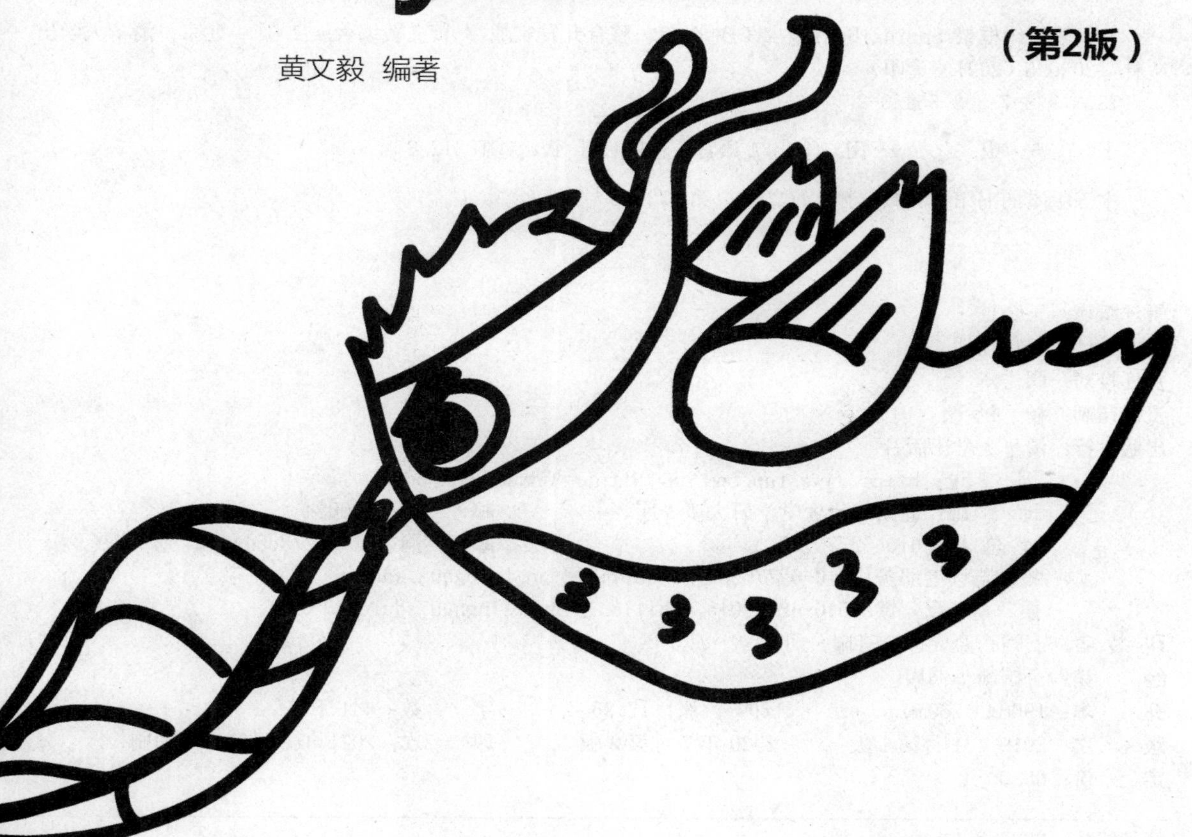

清華大學出版社
北 京

内 容 简 介

本书以 Spring 5、Spring MVC 5 和 MyBatis 3.X 为基础，从开发实战出发，结合最新的开发工具 IntelliJ IDEA，通过完整的项目实例让读者了解和学习 SSM 框架，又好又快地掌握 SSM 的开发技能。全书共分 11 章，第 1～2 章，由零开始，引导读者快速搭建 SSM 框架。第 3 章主要介绍 Spring 框架基础知识 IOC 和 AOP。第 4 章主要介绍 MyBatis 的映射器、动态 SQL、注解配置和关联映射。第 5 章主要介绍 Spring MVC 请求映射、参数绑定注解和信息转换详解。第 6 章主要介绍 MyBatis 的分页和分页插件 PageHelper、Spring 数据校验和 Spring 与 MyBatis 事务管理。第 7 章主要介绍 MyBatis 的一级缓存和二级缓存机制。第 8 章主要介绍 Spring MVC 执行流程、处理映射器和适配器以及视图解析器。第 9 章主要介绍 MyBatis 的整体框架、初始化流程和执行流程。第 10 章和第 11 章介绍用户管理系统和一个完整的高并发点赞项目的开发。

本书来自于一线开发人员的编程实践，突出技术的先进性和实用性，适用于所有 Java 编程语言开发人员、SSM 框架开发人员使用，也可作为培训机构和大专院校相关专业的教学用书。

图书在版编目（CIP）数据

Web 轻量级框架 Spring+Spring MVC+MyBatis 整合开发实战 / 黄文毅编著.—2 版.—北京：清华大学出版社，2020.5（2021.8重印）

ISBN 978-7-302-55420-2

Ⅰ. ①W… Ⅱ. ①黄… Ⅲ. ①JAVA 语言－程序设计 Ⅳ. ①TP312.8

中国版本图书馆 CIP 数据核字（2020）第 073318 号

责任编辑：王金柱
封面设计：王　翔
责任校对：闫秀华
责任印制：杨　艳
出版发行：清华大学出版社
　　网　　址：http://www.tup.com.cn，http://www.wqbook.com
　　地　　址：北京清华大学学研大厦 A 座　　**邮　　编：**100084
　　社 总 机：010-62770175　　**邮　　购：**010-62786544
　　投稿与读者服务：010-62776969，c-service@tup.tsinghua.edu.cn
　　质 量 反 馈：010-62772015，zhiliang@tup.tsinghua.edu.cn
印 装 者：三河市金元印装有限公司
经　　销：全国新华书店
开　　本：190mm×260mm　　**印　　张：**17.25　　**字　　数：**441 千字
版　　次：2019 年 1 月第 1 版　　2020 年 7 月第 2 版　　**印　　次：**2021 年 8 月第 2 次印刷
定　　价：68.00 元

产品编号：088757-01

前　言

Spring + Spring MVC + MyBatis（简称：SSM 框架）在 Java Web 开发领域中占据着十分重要的地位，一路走来已十余载，作为目前流行的轻量级 J2EE 框架，其在保留了经典 Java EE 应用架构高度可扩展性和高度可维护性的基础上，降低了 Java EE 应用的技术和部署成本，对于大部分企业应用是第一首选。因此掌握并学会使用 SSM 框架进行项目开发，成为 Java Web 开发人员必备技能之一。

本书以 Spring 5、Spring MVC 5 和 MyBatis 3.X 为基础，从开发实战出发，结合开发工具 IntelliJ IDEA，通过完整的项目实例让读者了解和学习 SSM 框架，又好又快地掌握 SSM 框架的开发技能。

本书结构

本书共 11 章，第 1 章至第 7 章主要是 SSM 框架基础知识篇，第 8 章和第 9 章主要是 Spring MCV 和 MyBatis 内部原理篇，第 10 和第 11 章项目实战篇。以下是各章的内容概要：

第 1 章主要介绍开始学习 SSM 框架之前的环境准备，包括 JDK 安装、IntelliJ IDEA 安装、Tomcat 安装和配置、Maven 安装以及 MySQL 数据库安装等。

第 2 章主要讲解如何一步一步快速搭建第一个 SSM 项目。

第 3 章主要回顾了 Spring 的基础知识 IOC 和 AOP、IOC 和 AOP 背后的实现原理以及设计模式。设计模式包括单例模式、简单工厂模式、工厂方法模式、动态代理模式等。

第 4 章主要介绍 MyBatis 常用的映射器元素、动态 SQL 元素、MyBatis 注解配置和关联映射。

第 5 章主要介绍 Spring MVC 常用注解，包括请求映射注解和参数绑定注解、Spring MVC 信息转换原理。

第 6 章主要介绍 MyBatis 提供的 RowBounds 分页的使用和原理，以及分页插件 PageHelper 的使用和原理，Spring 的 Validation 校验框架、JSR 303 校验、常用注解以及 Spring 和 MyBatis 事务管理。

第 7 章主要介绍 MyBatis 缓存机制，包括一级缓存和二级缓存以及一级缓存和二级缓存的使用及原理。

第 8 章主要介绍 Spring MVC 执行流程的原理剖析、前端控制器 DispatcherServlet 的原理、处理映射器和适配器的原理、视图解析器的原理等。

第 9 章主要介绍 MyBatis 整体框架、MyBatis 初始化流程及原理、MyBatis 执行流程及原理等。

第10章讲解一个用户管理系统的开发项目。

第11章主要介绍高并发项目常规解决方案，Redis缓存和消息中间件MQ的安装和使用以及如何一步一步实现高并发点赞项目。

学习本书的预备知识

Java基础

读者需要掌握J2SE基础知识，这是最基本的也是最重要的。

Java Web开发技术

在项目实战中需要用到Java Web的相关技术，比如HTML、Tomcat等技术。

数据库基础

读者需要掌握主流数据库基本知识，比如MySQL，同时掌握基本的SQL语法以及常用数据库的安装。

本书使用的软件版本

本书使用的开发环境为：

- 操作系统Windows 10
- 开发工具IntelliJ IDEA 2018.1
- JDK 1.8版本
- Tomcat 1.8版本
- Spring 5.0.4.RELEASE
- Spring MVC 5.0.4.RELEASE
- MyBatis 3.4.6

读者对象

本书适合所有Java编程语言开发人员，所有对Spring + Spring MVC + MyBatis感兴趣并希望使用SSM框架进行开发的人员，缺少SSM框架项目实战经验以及对SSM框架内部原理感兴趣的开发人员。

源代码、视频教学与PPT下载

GitHub源代码下载地址：

源码地址：https://github.com/huangwenyi10/springmvc-mybatis-book.git

配套视频地址：https://github.com/huangwenyi10/ssm-project.git

也可扫描二维码下载源代码：

扫描下面二维码下载教学视频：

01　　02

03　　04

扫描下面二维码下载 PPT：

如有下载问题，可发送电子邮件至 booksaga@126.com 获得帮助，邮件标题为“Web 轻量级框架 Spring+Spring MVC+MyBatis 整合开发实战（第 2 版）”。

致谢

本书能够顺利出版，首先要感谢清华大学出版社王金柱编辑给予分享技术、交流学习的机会，以及在本书出版过程的辛勤付出。

感谢厦门美图之家科技有限公司，书中很多的知识点和项目实战经验都来源于贵公司，感谢主管黄及峰、导师阮龙生和吴超群，同事林智泓、张汉铮、邱宗铭、尹权韬，项目管理王睿等在学习和生活上对我的照顾。

感谢家人，他们对我生活的照顾使得我没有后顾之忧，全身心投入到本书的写作当中。

限于水平和写作时间有限，欢迎大家通过电子邮件等方式批评指正。

编者

2020 年 3 月

目　录

第 1 章

准备 SSM 开发环境

本章首先简要介绍 SSM 框架，然后介绍开发环境的搭建，包括 JDK 的安装、IntelliJ IDEA 的安装、Tomcat 的安装和配置、Maven 的安装以及 MySQL 数据库的安装等内容。

1.1 SSM 简述

1.1.1 Spring 简述

Spring 开源框架是一个轻量级的企业级开发的一站式解决方案，是为了解决企业应用程序开发复杂性而创建的。基于 Spring 框架可以解决 Java EE 开发的所有问题。

Spring 框架是一个分层架构，由多个定义良好的模块组成，具体如图 1-1 所示。分层架构允许用户选择使用哪一个组件，同时为 J2EE 应用程序开发提供集成的框架。

1. Data Access/Integration（数据访问/集成）

- JDBC模块：提供了一个JBDC的样例模板，使用这些模板能消除传统冗长的JDBC编码和必须的事务控制，而且还能享受到Spring管理事务的好处。
- ORM模块：提供与流行的“对象/关系”映射框架的无缝集成，包括Hibernate、JPA、Ibatis等；而且可以使用Spring事务管理，无须额外控制事务。
- OXM模块：提供了一个Object/XML映射实现，将Java对象映射成XML数据，或者将XML数据映射成Java对象，Object/XML映射实现包括JAXB、Castor、XMLBeans和XStream等。

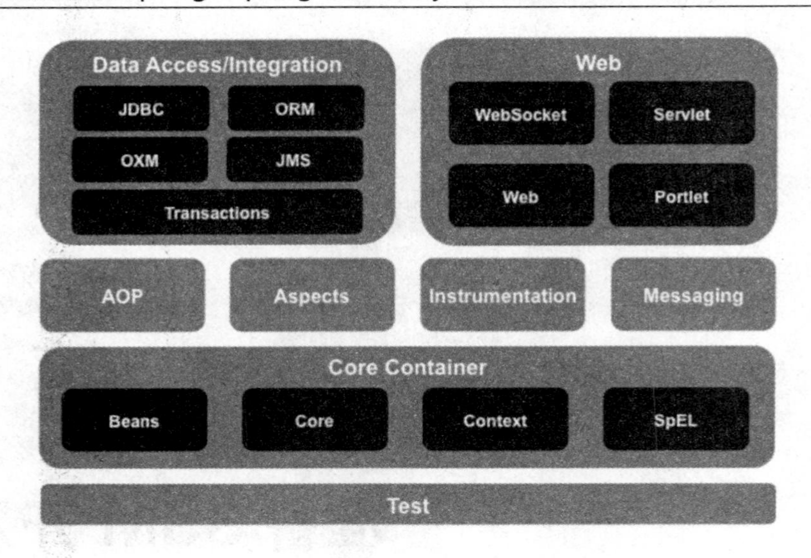

图 1-1　Spring 框架的分层结构

- **JMS模块**：提供一套“消息生产者、消息消费者”模板，使之更加简单地使用JMS。JMS用于在两个应用程序之间，或分布式系统中发送消息，进行异步通信。
- **Transactions模块**：该模块用于Spring管理事务，只要是Spring管理对象都能得到Spring管理事务的好处，无须在代码中进行事务控制了，而且支持编程和声明性的事务管理。

2. Web

- **WebSocket模块**：提供WebSocket功能。
- **Servlet模块**：提供了一个Spring MVC Web框架实现。Spring MVC框架提供了基于注解的请求资源注入、更简单的数据绑定、数据验证等以及一套非常易用的JSP标签，完全无缝与Spring其他技术协作。
- **Web模块**：提供了基础的Web功能。例如，多文件上传、集成IOC容器、远程过程访问（RMI、Hessian、Burlap）以及Web Service支持，并提供一个RestTemplate类来进行方便的Restful Services访问。
- **Portlet模块**：提供Portlet环境支持。

3. AOP、Aspects

- **AOP**：提供了符合 AOP Alliance规范的面向切面的编程（aspect-oriented programming）实现，提供比如日志记录、权限控制、性能统计等通用功能和业务逻辑分离的技术，并且能动态地把这些功能添加到需要的代码中。这样各司其职，降低了业务逻辑和通用功能的耦合。
- **Aspects**：提供了对AspectJ的集成，AspectJ提供了比Spring ASP更强大的功能。

4. Core Container（核心容器）

- **Spring-Beans**：提供了框架的基础部分，包括控制反转和依赖注入。其中Bean

Factory是容器核心，本质是“工厂设计模式”的实现，而且无须编程实现“单例设计模式”，单例完全由容器控制，而且提倡面向接口编程而非面向实现编程。所有应用程序对象及对象间的关系由框架管理，从而真正从程序逻辑中，把维护对象之间的依赖关系提取出来，所有这些依赖关系都由Bean Factory来维护。

- Spring-Core：核心工具类，封装了框架依赖的最底层部分，包括资源访问、类型转换及一些常用工具类。
- Spring-Context：以Core和Beans为基础，集成Beans模块功能并添加资源绑定、数据验证、国际化、Java EE支持、容器生命周期、事件传播等。核心接口是ApplicationContext。
- Spring-SpEL：提供强大的表达式语言支持，支持访问和修改属性值、方法调用；支持访问及修改数组、容器和索引器，命名变量；支持算数和逻辑运算；支持从Spring 容器获取Bean；也支持列表投影、选择和一般的列表聚合等。

5. Test

- Test模块：Spring支持JUNIT和TestNG测试框架，而且还额外提供了一些基于Spring的测试功能，比如在测试Web框架时，模拟HTTP请求的功能。

1.1.2 Spring MVC 简述

Spring MVC 属于 Spring FrameWork 的后续产品，已经融合在 Spring Web Flow 里面。Spring 框架提供了构建 Web 应用程序的全功能 MVC 模块。使用 Spring 可插入的 MVC 架构，从而在使用 Spring 进行 Web 开发时，可以选择使用 Spring 的 Spring MVC 框架或集成其他 MVC 开发框架，如 Struts1（现在一般不用）和 Struts2（一般老项目使用）等。

1.1.3 MyBatis 简述

MyBatis 本是 Apache 的一个开源项目 iBatis，2010 年这个项目由 Apache software foundation 迁移到了 google code，并且改名为 MyBatis。2013 年 11 月迁移到 Github。iBatis 一词来源于“internet”和“abatis”的组合，是一个基于 Java 的持久层框架。iBatis 提供的持久层框架包括 SQL Maps 和 Data Access Objects（DAOs）。

MyBatis 是一款优秀的持久层框架，它支持定制化 SQL、存储过程及高级映射。MyBatis 避免了几乎所有的 JDBC 代码和手动设置参数以及获取结果集。MyBatis 可以使用简单的 XML 或注解来配置和映射原生信息，将接口和 Java 的 POJOs 映射成数据库中的记录。

以上我们对 SSM 框架做了一个整体介绍，接下来将学习如何搭建 Spring、Spring MVC 和 MyBatis 的开发环境，以便读者可以进行项目开发，包括 JDK 的安装、IntelliJ IDEA 的安装、Tomcat 的安装和配置、Maven 的安装以及 MySQL 数据库的安装等内容。

1.2 JDK 安装

JDK（Java Development Kit）是 Java 语言的软件开发工具包，由 SUN 公司提供。JDK 是整个 Java 开发的核心，它包含了 Java 的运行环境（JVM + Java 系统类库）和 Java 工具，所有 Java 程序的编写都依赖于它。

下面介绍 JDK 的安装。

JDK 建议使用 1.8 及以上的版本，其官方下载路径为：http://www.oracle.com/technetwork/java/javase/downloads/jdk8-downloads-2133151.html。大家可以根据自己 Windows 操作系统的配置选择合适的 JDK 1.8 安装包，这里就不过多描述。

软件下载完成之后，双击下载软件，出现安装界面，如图 1-2 所示。一直单击【下一步】按钮，即可完成安装。这里笔者把 JDK 安装在路径 C:\Program Files\Java\jdk1.8.0_77 下。

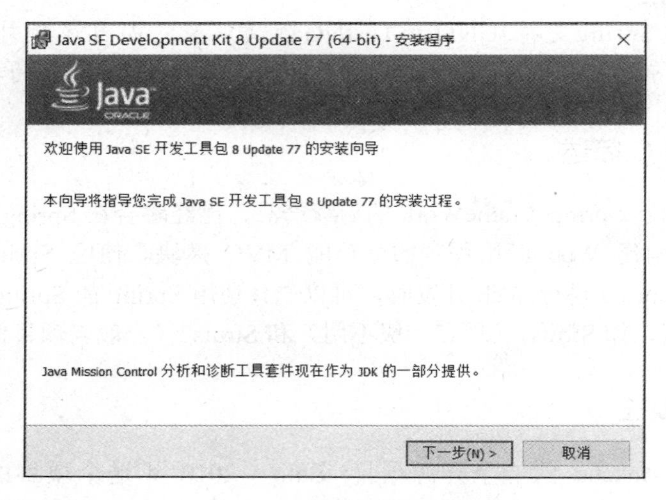

图 1-2　JDK 安装界面

安装完成后，需要配置环境变量 JAVA_HOME，具体步骤如下：

步骤 01 在电脑桌面上，右击【我的电脑】→【属性】→【高级系统设置】→【环境变量】→【系统变量(S)】→【新建】，出现新建环境变量的窗口，如图 1-3 所示。

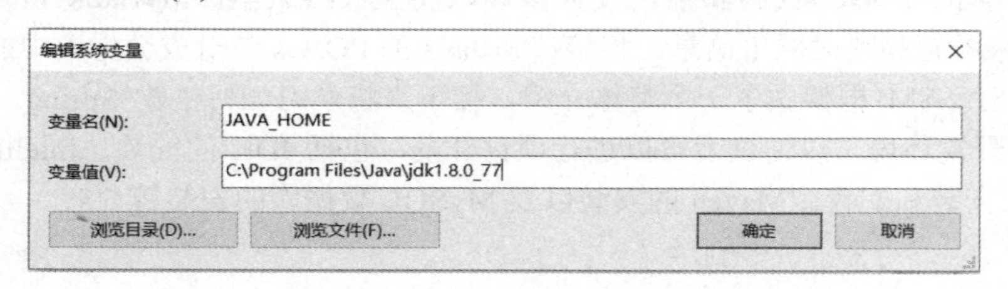

图 1-3　新建环境变量窗口

步骤 02 在【变量名】和【变量值】文本框中分别填入 JAVA_HOME 和 C:\Program Files\Java\ jdk1.8.0_77，单击【确定】按钮。

步骤 03 JAVA_HOME 配置好之后，将%JAVA_HOME%\bin 加入到【系统变量】的 path 中。配置完成之后，打开命令行窗口，输入命令 java -version。出现如图 1-4 所示的提示，即表示安装成功。

```
命令提示符
Microsoft Windows [版本 10.0.14393]
(c) 2016 Microsoft Corporation。保留所有权利。

C:\Users\Ay>java -version
java version "1.8.0_77"
Java(TM) SE Runtime Environment (build 1.8.0_77-b03)
Java HotSpot(TM) 64-Bit Server VM (build 25.77-b03, mixed mode)

C:\Users\Ay>_
```

图 1-4　安装成功的命令行窗口

JDK 安装路径最好不要出现中文，否则会出现意想不到的错误。

1.3　IntelliJ IDEA 安装

IDEA 全称为 IntelliJ IDEA，是 Java 语言开发的集成环境，IntelliJ 在业界被公认为最好的 Java 开发工具之一，尤其在智能代码助手、代码自动提示、重构、J2EE 支持、各类版本工具（Git、Svn、Github 等）、JUnit、CVS 整合、代码分析、创新的 GUI 设计等方面的功能可以说是超常的。IDEA 是 JetBrains 公司的产品，这家公司总部位于捷克共和国的首都布拉格，开发人员以严谨著称的东欧程序员为主。它的旗舰版本还支持 HTML、CSS、PHP、MySQL、Python 等。免费版只支持 Java 等少数语言。

如果你还在使用 Eclipse 或者 MyEclipse 等开发工具进行代码开发，强烈建议大家切换至 IntelliJ IDEA 开发工具。目前所有大型的互联网公司，比如百度、腾讯、阿里、美团等，都是使用 IntelliJ IDEA 进行项目开发的，IDEA 是目前的主流开发工具，会极大地提高你的开发效率。

在 IntelliJ IDEA 的官方网站 http://www.jetbrains.com/idea/可以免费下载 IDEA。下载完 IDEA 后，运行安装程序，按提示安装即可。本书使用 IntelliJ IDEA 2016.2 版本，当然大家也可以使用其他版本的 IDEA，只要版本不要过低即可。安装成功之后，软件界面如图 1-5 所示。

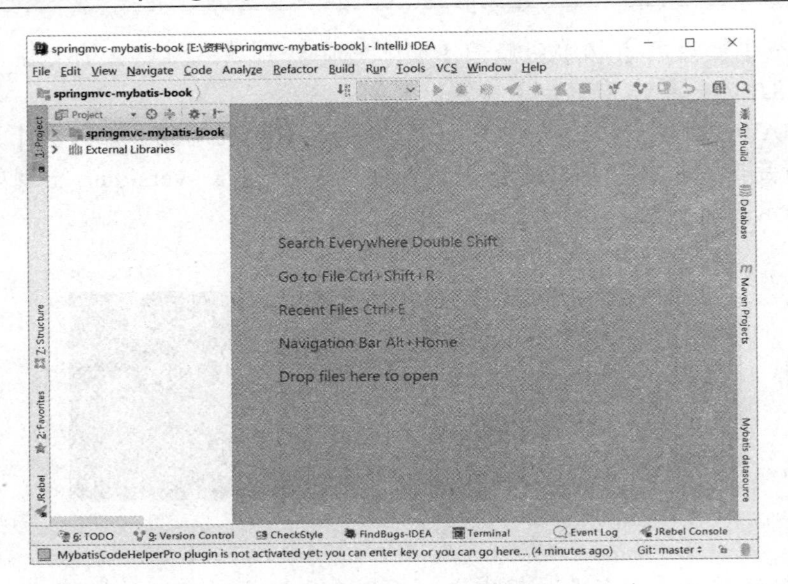

图 1-5　IntelliJ IDEA 软件窗口

1.4　Tomcat 的安装与配置

Tomcat 服务器是一个免费的开放源代码的 Web 应用服务器，属于轻量级应用服务器。因为 Tomcat 技术先进、性能稳定，而且免费，因而深受 Java 爱好者的喜爱并得到了部分软件开发商的认可，成为目前比较流行的 Web 应用服务器。

1.4.1　Tomcat 的下载

本书使用 Tomcat 8.0 进行讲解，可到官网 https://tomcat.apache.org/download-80.cgi 进行下载，下载完成之后解压到 D：盘，并将解压后的文件夹命名为 tomcat8，具体如图 1-6 所示。

本地磁盘 (D:) > tomcat8 >

名称	修改日期	类型	大小
bin	2017/12/18 22:21	文件夹	
conf	2017/12/18 22:21	文件夹	
lib	2017/12/18 22:21	文件夹	
logs	2016/3/18 20:31	文件夹	
temp	2017/12/18 22:21	文件夹	
webapps	2017/12/18 22:21	文件夹	
work	2016/3/18 20:31	文件夹	
LICENSE	2016/3/18 20:32	文件	57 KB
NOTICE	2016/3/18 20:32	文件	2 KB
RELEASE-NOTES	2016/3/18 20:32	文件	7 KB
RUNNING.txt	2016/3/18 20:32	文本文档	17 KB

图 1-6　Tomcat 解压目录

1.4.2　IntelliJ IDEA 配置 Tomcat

在 IntelliJ IDEA 中配置 Tomcat，具体步骤如下：

步骤 01　在 IDEA 开发菜单栏中，选择【run】→【Edit Configurations】，在弹出的窗口中选择【Defaults】→【Tomcat Server】→【Local】，在【Application server】中选择 Tomcat 的安装路径，在【JRE】中选择 JDK 的安装路径，最后单击【Apply】→【OK】确认，具体如图 1-7 所示。

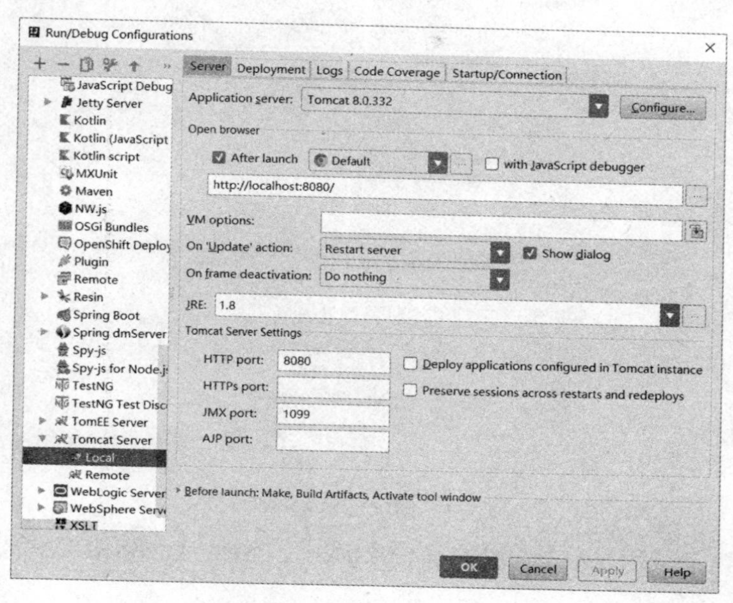

图 1-7　Tomcat 配置

步骤 02　步骤一只是配置一个 Defaults 默认 Tomcat 模板，现在我们单击【+】加号按钮→【Tomcat Server】→【Local】，在弹出的界面中输入 Name 为 tomcat8，其他信息会从默认模板中获取到，具体如图 1-8 和图 1-9 所示。

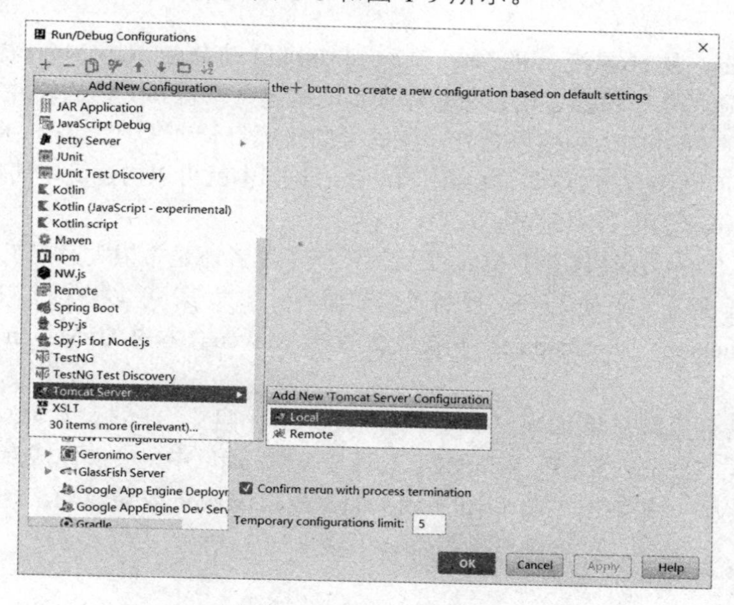

图 1-8　创建 Tomcat 配置

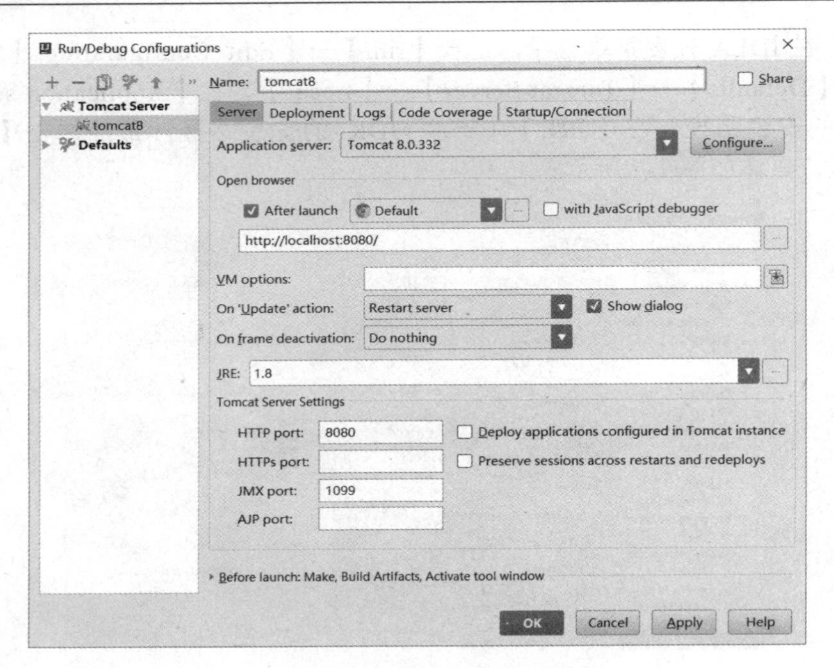

图 1-9　修改 Tomcat 名称

步骤 03　设置完成后，单击【Apply】→【OK】。至此，IntelliJ IDEA 配置 Tomcat 大功告成。

1.5　Maven 的安装和配置

Apache Maven 是目前流行的项目管理和构建自动化工具。Maven 项目对象模型（POM），可以通过一小段描述信息来管理项目的构建、报告和文档的软件项目管理工具。Maven 除了以程序构建能力为特色之外，还提供高级项目管理工具。由于 Maven 的默认构建规则有较高的可重用性，所以常常用两三行 Maven 脚本就可以构建简单的项目。

下面介绍 Maven 的安装和配置。

虽然 IntelliJ IDEA 已经包含了 Maven 插件，但是笔者还是希望大家在工作中能够安装自己的 Maven 插件，方便以后项目配置需要。大家可以通过 Maven 的官网 http://maven.apache.org/ download.cgi 下载最新版的 Maven。本书的 Maven 版本为 apache-maven-3.5.0。

Maven 下载完后解压缩即可。例如，解压到 D：盘上，然后将 Maven 的安装路径 D:\apache-maven-3.5.0\bin 加入到 Windows 的环境变量 path 中。安装完成后，在命令行窗口执行命令：mvn -v，如果输出如图 1-10 所示页面，表示 Maven 安装成功。

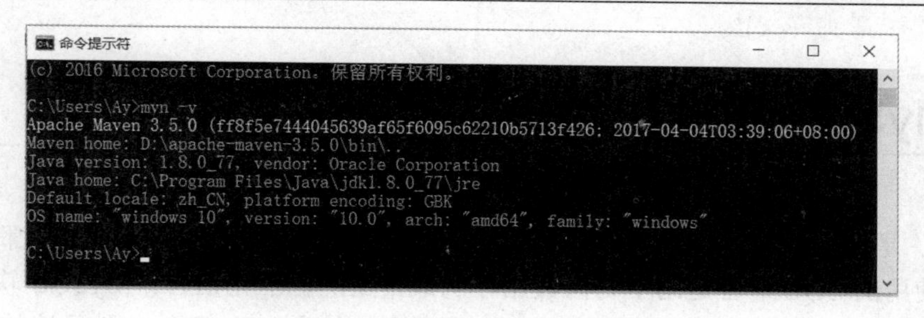

图 1-10　Maven 安装成功命令行窗口

接下来，我们在 IntelliJ IDEA 下配置 Maven，具体步骤如下：

步骤 01 在 Maven 安装目录，即 D:\apache-maven-3.5.0 下新建文件夹 repository，用来作为本地仓库。

步骤 02 在 IntelliJ IDEA 界面中，选择【File】→【Settings】，在出现的窗口中找到 Maven 选项，分别把【Maven home directory】【User settings file】【Local repository】，设置为我们自己 Maven 的相关目录，如图 1-11 所示。

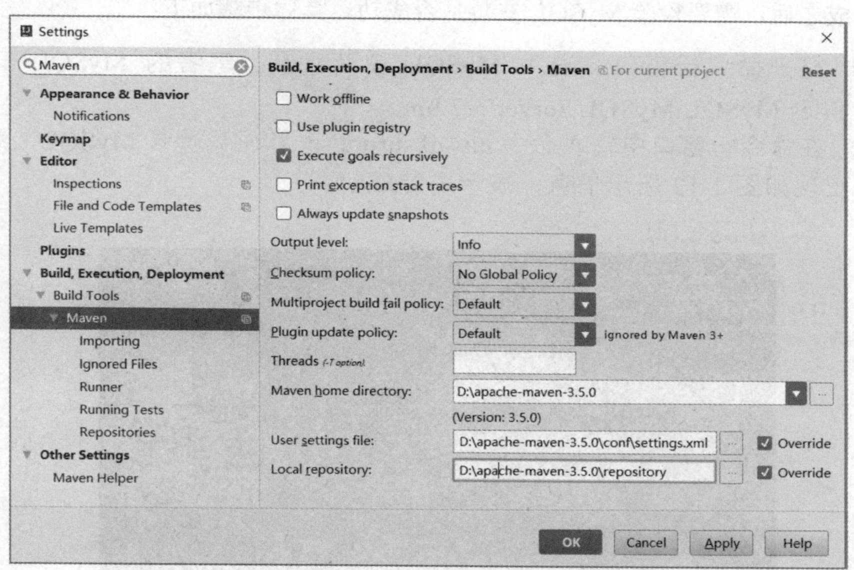

图 1-11　Maven 设置窗口

步骤 03 设置完成后，单击【Apply】→【OK】。至此，Maven 在 IntelliJ IDEA 的配置完成。

注意

之所以把 Maven 默认仓库（C:\${user.home}\.m2\respository）的路径改为我们自己的目录（D:\apache-maven-3.5.0\repository），是因为 respository 仓库到时候会存放很多的 jar 包，放在 C: 盘影响计算机的性能，所以才会修改默认仓库的位置。

1.6 MySQL 数据库的安装

MySQL 是目前项目中运用广泛的关系型数据库，无论在什么样的公司都运用甚广。MySQL 所使用的 SQL 语言是用于访问数据库的最常用的标准化语言。MySQL 软件由于体积小、速度快、总体拥有成本低，尤其是开发源码这一特点，一般中小型网站的开发都选择 MySQL 作为网站数据库。

1.6.1 MySQL 的安装

MySQL 的安装很简单，安装方式也有多种。大家可以到 MySQL 的官网 https://dev.mysql.com/ downloads/mysql/下载 MySQL 安装软件，并按照提示一步一步安装即可。如果你的计算机已经安装了 MySQL，可略过此节。本书使用的 MySQL 版本为 5.7.17。

安装完成之后，需要检验 MySQL 安装是否成功。具体步骤如下：

步骤 01 打开命令行窗口，进入 MySQL 安装目录，笔者的 MySQL 安装目录是 C:\Program Files\MySQL\MySQL Server 5.7\bin。

步骤 02 在命令行窗口中输入命令 mysql -uroot -p 和密码登录 MySQL，然后再输入命令 status 出现如图 1-12 所示信息，表示安装成功。

```
mysql -uroot -p

C:\Program Files\MySQL\MySQL Server 5.7\bin>mysql -uroot -p
Enter password: ******
Welcome to the MySQL monitor.  Commands end with ; or \g.
Your MySQL connection id is 12
Server version: 5.7.17-log MySQL Community Server (GPL)

Copyright (c) 2000, 2016, Oracle and/or its affiliates. All rights reserved.

Oracle is a registered trademark of Oracle Corporation and/or its
affiliates. Other names may be trademarks of their respective
owners.

Type 'help;' or '\h' for help. Type '\c' to clear the current input statement.

mysql> status;
--------------
mysql  Ver 14.14 Distrib 5.7.17, for Win64 (x86_64)

Connection id:          12
Current database:
Current user:           root@localhost
SSL:                    Not in use
Using delimiter:        ;
Server version:         5.7.17-log MySQL Community Server (GPL)
Protocol version:       10
Connection:             localhost via TCP/IP
Server characterset:    utf8
Db     characterset:    utf8
Client characterset:    gbk
```

图 1-12　MySQL 安装状态

1.6.2 Navicat for MySQL 客户端安装与使用

Navicat for MySQL 是连接 MySQL 数据库的客户端工具，通过使用该客户端工具，方便我们对数据库进行操作，比如创建数据库表、添加数据等。如果大家已经安装了其他的

MySQL 客户端，可以略过本节。

Navicat for MySQL 的安装也非常简单，大家可以到网上下载安装即可。安装完成之后，打开软件，如图 1-13 所示。

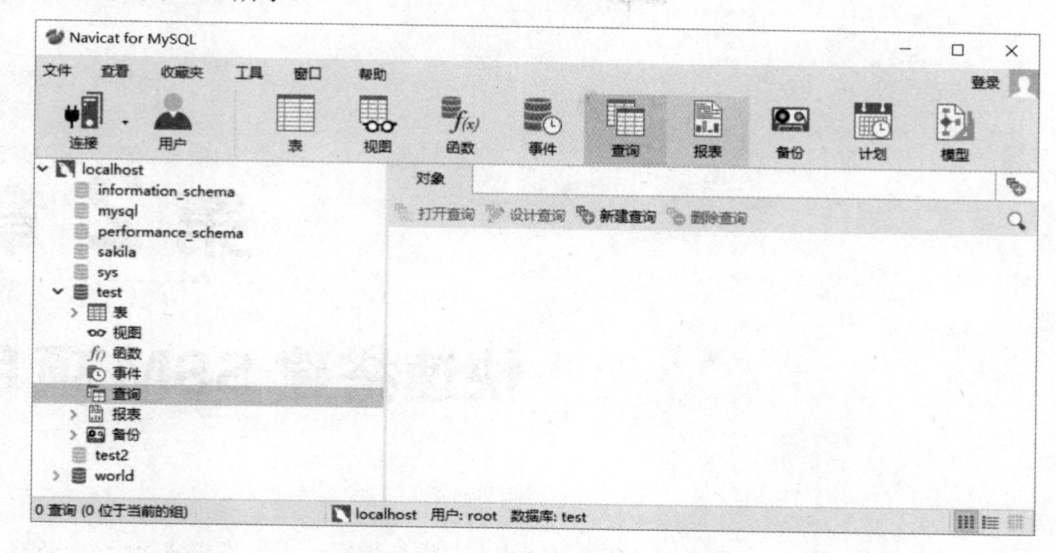

图 1-13　Navicat for MySQL 界面

可以通过【查询】→【新建查询】，在弹出的窗口中编写相关的 SQL 语句来查询数据。当然还有很多的操作，大家可以自己去使用和掌握它，这里就不一一描述了。

1.7　思考与练习

1. 简述什么是 Spring 框架？
2. Spring 框架由有哪些模块组成？
3. 简述什么是 Spring MVC 框架？
4. 什么是 JDK？
5. 如何在 Windows 操作系统下安装 JDK？
6. 什么是 Tomcat？
7. 什么是 Maven？
8. MySQL 是什么类型的数据库？MySQL 有哪些优点？
9. Navicat for MySQL 和 MySQL 是什么关系？

第 2 章

快速搭建 SSM 项目

本章将讲解使用 Spring、Spring MVC、MyBatis 框架如何一步一步搭建第一个 SSM 项目，以便读者能够快速掌握 SSM 框架开发项目的具体流程和步骤，从而为实际开发打下基础。

2.1 快速搭建 Web 项目

快速搭建 Web 项目的操作步骤如下：

步骤 01 在 IntelliJ IDEA 的菜单栏中选择【File】→【New】→【Project...】，在弹出的【New Project】窗口中选择【Maven】，勾选【Create from archetype】，选择【maven-archetype-webapp】选项，单击【Next】按钮，具体如图 2-1 所示。

步骤 02 在图 2-2 中，填写【GroupId】和【ArtifactId】等信息后，单击【Next】按钮。

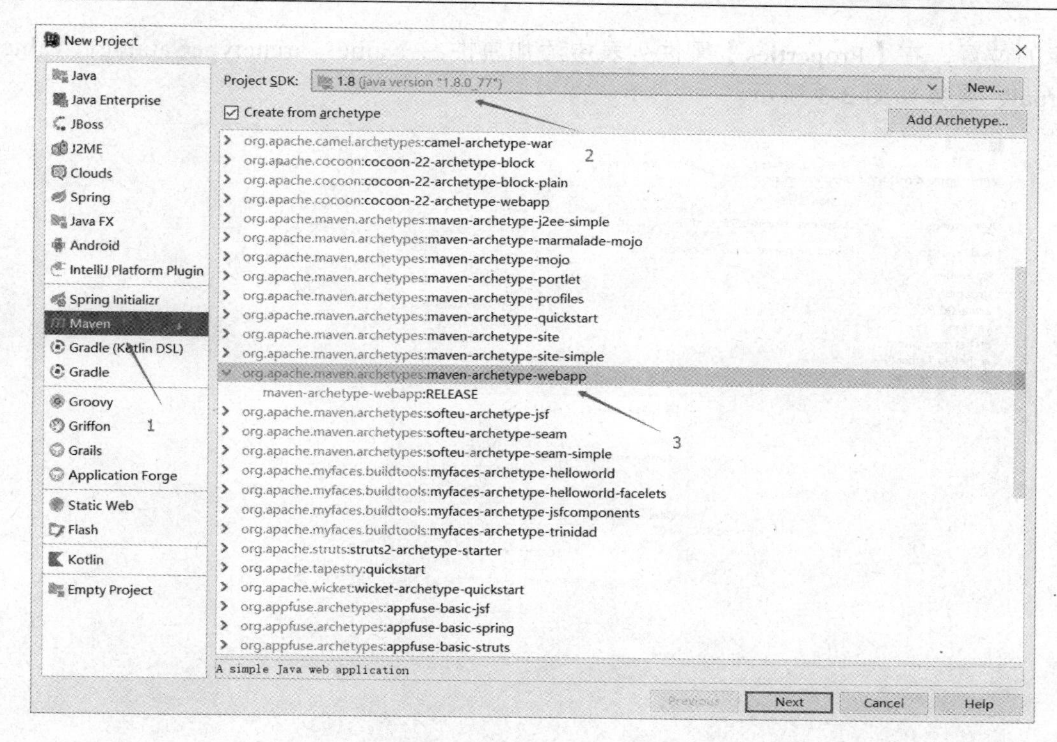

图 2-1 New Product 窗口

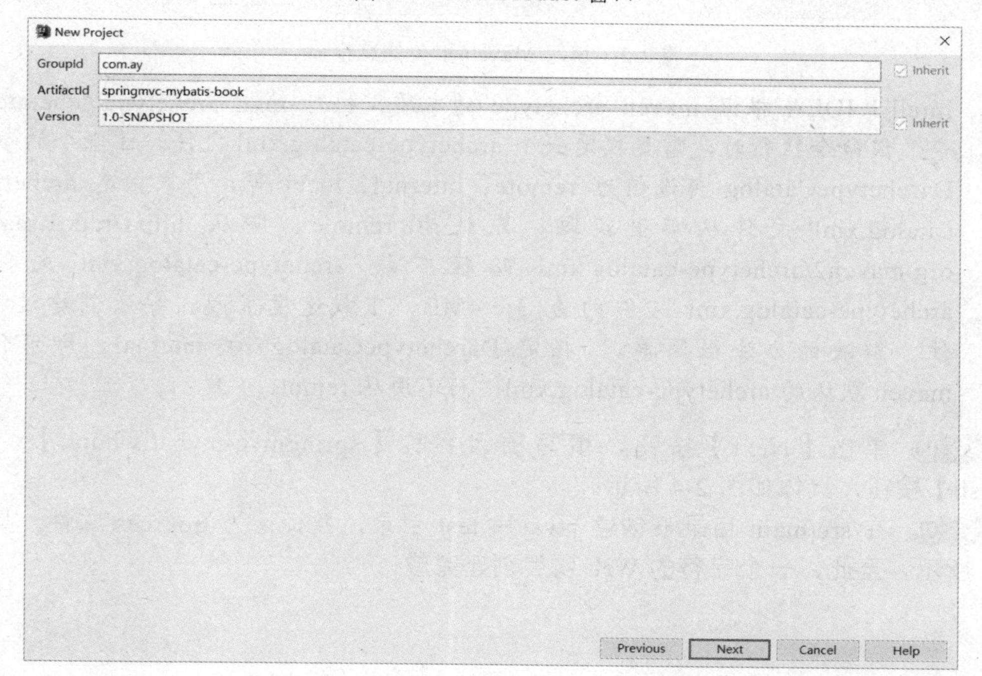

图 2-2 填写 Maven 相关信息窗口

步骤 03 在【Maven home directory】中选择 Maven 的安装路径，具体见 1.5 节 Maven 安装。在【User settings file】和【Local repository】中选择 Maven 的配置文件和

仓库的位置，在【Properties】属性列表中添加属性名 Name：archetypeCatalog；Value：internal，具体如图 2-3 所示。

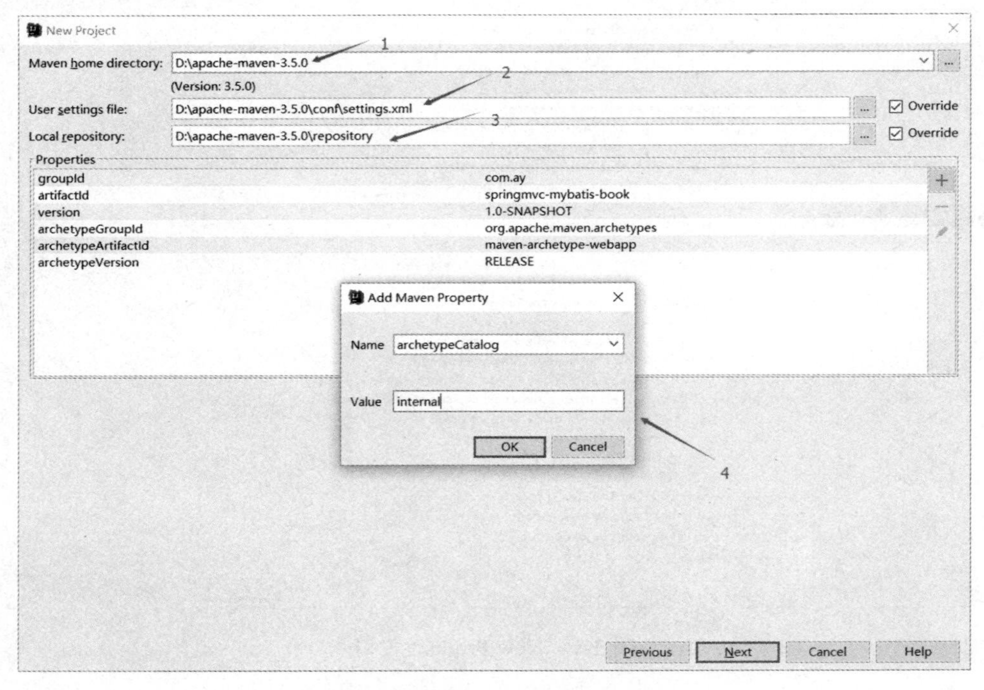

图 2-3　填写 Maven 相关信息窗口

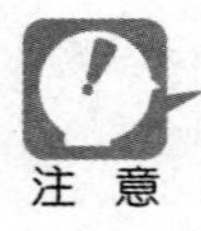

注意

IntelliJ IDEA 根据 maven archetype 的本质，执行 mvn archetype:generate 命令。该命令执行时，需要指定一个 archetype-catalog.xml 文件。该命令的参数-DarchetypeCatalog 可选值为 remote、internal、local 等，用来指定 archetype-catalog.xml 文件从哪里获取，默认为 remote，即从 http://repo1.maven.org/maven2/archetype-catalog.xml 路径下载 archetype-catalog.xml 文件。archetype-catalog.xml 文件约为 3～4MB，下载速度很慢，导致创建过程卡住。解决的办法很简单，指定-DarchetypeCatalog 为 internal，即可使用 maven 默认的 archetype-catalog.xml，而不用从 remote 下载。

步骤 04 单击【Next】按钮，填写项目名称【springmvc-mybatis-book】，单击【Finish】按钮，具体如图 2-4 所示。

步骤 05 在/src/main 目录下创建 java 和 test 目录，并标记为 Sources 文件，具体如图 2-5 所示。至此，一个完整的 Web 项目创建完成。

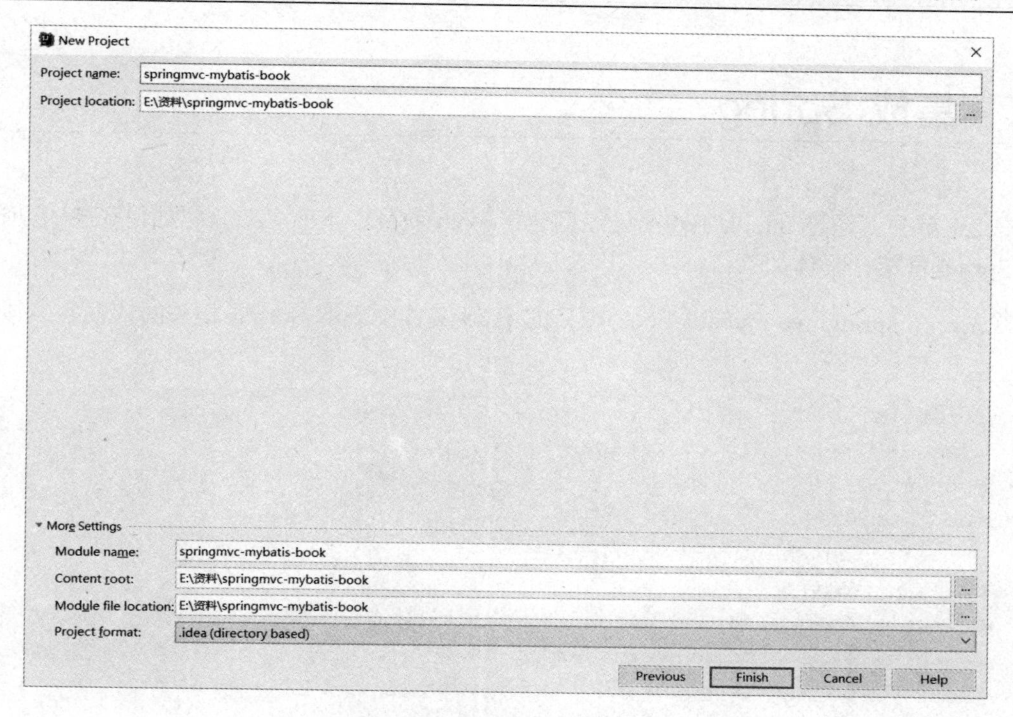

图 2-4　填写项目相关信息

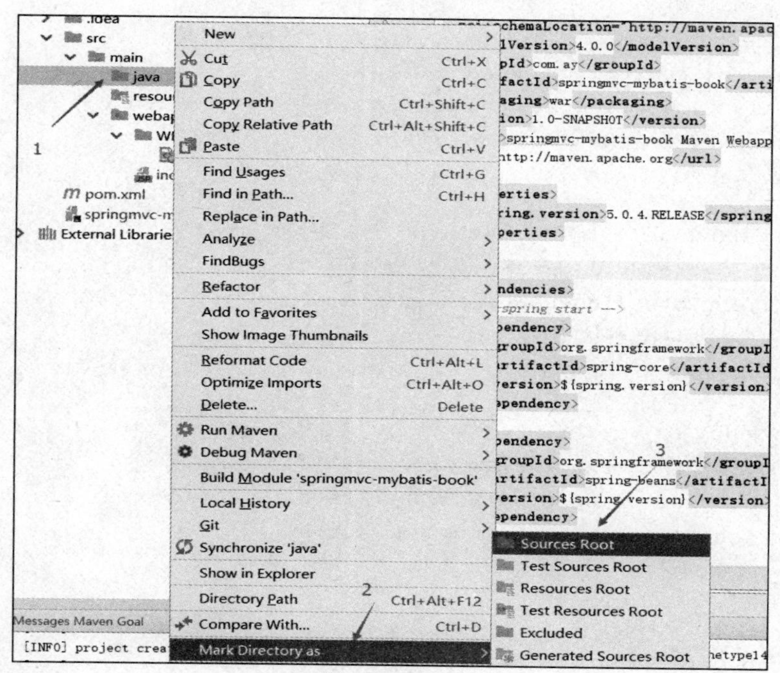

图 2-5　创建 java 目录

2.2 集成 Spring

在 2.1 节中，通过 IntelliJ IDEA 已经创建好 Web 项目，本节主要介绍如何在 Web 项目中集成 Spring 框架，具体如下：

首先，在 springmvc-mybatis-book 项目的 pom 文件中添加 Spring 相关的依赖，具体代码如下：

```
<properties>
    <spring.version>5.0.4.RELEASE</spring.version>
</properties>
<dependencies>
    <!--spring start -->
    <dependency>
      <groupId>org.springframework</groupId>
      <artifactId>spring-core</artifactId>
      <version>${spring.version}</version>
    </dependency>

    <dependency>
      <groupId>org.springframework</groupId>
      <artifactId>spring-beans</artifactId>
      <version>${spring.version}</version>
    </dependency>

    <dependency>
      <groupId>org.springframework</groupId>
      <artifactId>spring-context</artifactId>
      <version>${spring.version}</version>
    </dependency>

    <dependency>
      <groupId>org.springframework</groupId>
      <artifactId>spring-context-support</artifactId>
      <version>${spring.version}</version>
    </dependency>

    <dependency>
      <groupId>org.springframework</groupId>
      <artifactId>spring-aop</artifactId>
      <version>${spring.version}</version>
    </dependency>

    <dependency>
```

```
        <groupId>org.springframework</groupId>
        <artifactId>spring-aspects</artifactId>
        <version>${spring.version}</version>
      </dependency>

      <dependency>
        <groupId>org.springframework</groupId>
        <artifactId>spring-expression</artifactId>
        <version>${spring.version}</version>
      </dependency>

      <dependency>
        <groupId>org.springframework</groupId>
        <artifactId>spring-tx</artifactId>
        <version>${spring.version}</version>
      </dependency>

      <dependency>
        <groupId>org.springframework</groupId>
        <artifactId>spring-test</artifactId>
        <version>${spring.version}</version>
      </dependency>

     <dependency>
        <groupId>org.springframework</groupId>
        <artifactId>spring-web</artifactId>
        <version>${spring.version}</version>
      </dependency>
      <!--spring end -->

  <!-- junit -->
      <dependency>
        <groupId>junit</groupId>
        <artifactId>junit</artifactId>
        <version>4.12</version>
      </dependency>
  </dependencies>
```

其次，在/src/main 下创建 resources 目录并标记为 resources，在 resources 目录下创建 applicationContext.xml 配置文件，具体代码如下：

```
  <?xml version="1.0" encoding="UTF-8"?>
  <beans xmlns="http://www.springframework.org/schema/beans"
         xmlns:xsi="http://www.w3.org/2001/XMLSchema-instance"
         xmlns:context="http://www.springframework.org/schema/context"
         xmlns:tx="http://www.springframework.org/schema/tx"
         xsi:schemaLocation="http://www.springframework.org/schema/beans
```

```
            http://www.springframework.org/schema/beans/ spring-beans-2.5.xsd
            http://www.springframework.org/schema/tx
            http://www.springframework.org/schema/tx/spring-tx.xsd
            http://www.springframework.org/schema/context
            http://www.springframework.org/schema/context/spring-context-
2.5.xsd">
        <!—在/src/main/java 目录下创建 package:  com.ay  -->
        <context:component-scan base-package="com.ay"/>

    </beans>
```

- <context:component-scan/>注解：扫描base-package包或者子包下所有的Java类，并把匹配的Java类注册成Bean。这里我们设置扫描com.ay包下的所有Java类。

接着，在 web.xml 配置文件中添加如下代码：

```
<!DOCTYPE web-app PUBLIC
 "-//Sun Microsystems, Inc.//DTD Web Application 2.3//EN"
 "http://java.sun.com/dtd/web-app_2_3.dtd" >
<web-app>
  <display-name>Archetype Created Web Application</display-name>
  <context-param>
    <param-name>contextConfigLocation</param-name>
    <param-value>classpath:applicationContext.xml</param-value>
  </context-param>
  <listener>
  <listener-class>
org.springframework.web.context.ContextLoaderListener
</listener-class>
  </listener>
</web-app>
```

- <context-param>：整个项目的全局变量，相当于设定了一个固定值。param-name是键，相当于参数名，param-value是值，相当于参数值。
- ContextLoaderListener：ContextLoaderListener监听器实现了ServletContextListener接口。其作用是启动Web容器时，自动装配ApplicationContext的配置信息。在web.xml配置这个监听器，启动容器时，就会默认执行它实现的方法。

最后，在 src/main/test/com.ay.test 目录下创建 SpringTest 测试类，具体代码如下：

```
import org.junit.Test;
import org.springframework.context.ApplicationContext;
import org.springframework.context.support.ClassPathXmlApplicationContext;
import org.springframework.stereotype.Service;
/**
 * @author Ay
```

```
 * @date 2018/04/02
 */
@Service
public class SpringTest {

    @Test
    public void testSpring(){
        //获取运用上下文
        ApplicationContext applicationContext =
                new ClassPathXmlApplicationContext ("applicationContext.xml");
        //获取SpringTest类
        SpringTest springTest = (SpringTest) applicationContext.getBean
("springTest");
        //调用sayHello方法
springTest.sayHello();
    }

    public void sayHello(){
        System.out.println("hello ay");
    }
}
```

- @Service：Spring会自动扫描到@Service注解的类，并把这些类纳入进Spring容器中管理。也可以用@Component注解，只是@Service注解更能表明该类是服务层类。
- ApplicationContext容器：ApplicationContext是Spring中较高级的容器，它可以加载配置文件中定义的Bean，并将所有的Bean集中在一起，当有请求的时候分配Bean。

最经常被使用的ApplicationContext接口实现如下：

- ClassPathXmlApplicationContext：从类路径ClassPath中寻找指定的XML配置文件，找到并装载完成ApplicationContext的实例化工作，具体代码如下：

```
//装载单个配置文件实例化ApplicationContext容器
ApplicationContext cxt = new
ClassPathXmlApplicationContext("applicationContext.xml");
//装载多个配置文件实例化ApplicationContext容器
String[] configs = {"bean1.xml","bean2.xml","bean3.xml"};
ApplicationContext cxt = new ClassPathXmlApplicationContext(configs);
```

- FileSystemXmlApplicationContext：从指定的文件系统路径中寻找指定的XML配置文件，找到并装载完成ApplicationContext的实例化工作。具体代码如下：

```
//装载单个配置文件实例化ApplicationContext容器
ApplicationContext cxt = new FileSystemXMLApplicationContext("beans.xml");
//装载多个配置文件实例化ApplicationContext容器
```

```
String[] configs = {"c:/beans1.xml","c:/beans2.xml"};
ApplicationContext cxt = new FileSystemXmlApplicationContext(configs);
```

- XmlWebApplicationContext：从Web应用中寻找指定的XML配置文件，找到并装载完成ApplicationContext的实例化工作。这是为Web工程量身定制的，使用WebApplicationContextUtils类的getRequiredWebApplicationContext方法可在JSP与Servlet中取得IOC容器的引用。

运行上面代码中的单元测试方法 testSpring()，便可以在 IntelliJ IDEA 控制台看到如图 2-6 所示的结果，表示 Web 应用集成 Spring 框架成功。

```
Connected to the target VM, address: '127.0.0.1:65035', transport: 'socket'
四月 02, 2018 12:53:33 下午 org.springframework.context.support.AbstractApplicationContext prepareRefresh
信息: Refreshing org.springframework.context.support.ClassPathXmlApplicationContext@52525845: startup dat
四月 02, 2018 12:53:34 下午 org.springframework.beans.factory.xml.XmlBeanDefinitionReader loadBeanDefiniti
信息: Loading XML bean definitions from class path resource [applicationContext.xml]
hello ay
Disconnected from the target VM, address: '127.0.0.1:65035', transport: 'socket'

Process finished with exit code 0
```

图 2-6　Web 应用集成 Spring 框架

2.3 集成 Spring MVC 框架

Web 项目集成 Spring 框架之后，我们继续把 Spring MVC 集成进来，具体如下：

首先，把集成 Spring MVC 所需要的 Maven 依赖包和相关的属性值添加到 pom.xml 文件中，具体代码如下：

```
<properties>
   <spring.version>5.0.4.RELEASE</spring.version>
   <javax.servlet.version>4.0.0</javax.servlet.version>
   <jstl.version>1.2</jstl.version>
</properties>
<!--springmvc start -->
<dependency>
     <groupId>jstl</groupId>
     <artifactId>jstl</artifactId>
     <version>${jstl.version}</version>
</dependency>
<dependency>
     <groupId>javax.servlet</groupId>
     <artifactId>javax.servlet-api</artifactId>
     <version>${javax.servlet.version}</version>
</dependency>
```

```
<dependency>
    <groupId>org.springframework</groupId>
    <artifactId>spring-webmvc</artifactId>
    <version>${spring.version}</version>
</dependency>
<!--springmvc end -->
```

其次，在 web.xml 配置文件中添加 DispatcherServlet 配置，具体代码如下：

```
<!--配置 DispatcherServlet -->
<servlet>
  <servlet-name>spring-dispatcher</servlet-name>
  <servlet-class>org.springframework.web.servlet.DispatcherServlet
</servlet-class>
  <!-- 配置 SpringMVC 需要加载的配置文件 spring-mvc.xml -->
  <init-param>
    <param-name>contextConfigLocation</param-name>
    <param-value>classpath:spring-mvc.xml</param-value>
  </init-param>
  <load-on-startup>1</load-on-startup>
</servlet>
<servlet-mapping>
  <servlet-name>spring-dispatcher</servlet-name>
  <!-- 默认匹配所有的请求 -->
  <url-pattern>/</url-pattern>
</servlet-mapping>
```

- DispatcherServlet类：DispatcherServlet是前置控制器，主要用于拦截匹配的请求，拦截匹配规则要自己定义，把拦截下来的请求，依据相应的规则分发到目标Controller来处理，是配置Spring MVC的第一步。
- <init-param>：整个项目的局部变量，相当于设定了一个固定值，只能在当前的Servlet中被使用。param-name是键，相当于就是参数名，param-value是值，相当于参数值。容器启动时会加载配置文件spring-mvc.xml。
- <load-on-startup>：表示启动容器时初始化该Servlet。当值为0或者大于0时，表示容器在应用启动时加载并初始化这个Servlet。如果值小于0或未指定时，则指示容器在该Servlet被选择时才加载。正值越小，Servlet的优先级越高，应用启动时就越先加载。值相同时，容器就会自己选择顺序来加载。
- <servlet-mapping>：标签声明了与该Servlet相应的匹配规则，每个<url-pattern>标签代表一个匹配规则。
- <url-pattern>：URL匹配规则，表示哪些请求交给Spring MVC处理，“/”表示拦截所有的请求。

URL 匹配规则有如下几种：

（1）精准匹配

<url-pattern>中的配置项必须与 URL 完全精确匹配。

```
<servlet-mapping>
    <servlet-name>spring-dispatcher</servlet-name>
    <!-- 精准匹配 -->
    <url-pattern>/ay</url-pattern>
    <url-pattern>/index.html</url-pattern>
    <url-pattern>/test/ay.html</url-pattern>
</servlet-mapping>
```

当在浏览器中输入如下几种 URL 时，都会被匹配到该 Servlet：

```
http://localhost/ay
http://localhost/index.html
http://localhost/test/ay.html
```

（2）扩展名匹配

以“*.”开头的字符串被用于扩展名匹配。

```
<servlet-mapping>
    <servlet-name>spring-dispatcher</servlet-name>
    <!-- 扩展名匹配 -->
    <url-pattern>*.jsp</url-pattern>
</servlet-mapping>
```

当在浏览器中输入如下几种 URL 时，都会被匹配到该 Servlet，具体代码如下：

```
http://localhost/ay.jsp
http://localhost/al.jsp
```

（3）路径匹配

以“/”字符开头，并以“/*”结尾的字符串用于路径匹配。

```
<servlet-mapping>
    <servlet-name>spring-dispatcher</servlet-name>
    <!-- 扩展名匹配 -->
    <url-pattern>/ay/*</url-pattern>
</servlet-mapping>
```

当在浏览器中输入如下几种 URL 时，都会被匹配到该 Servlet：

```
http://localhost/ay/ay.jsp
http://localhost/ay/ay.html
http://localhost/ay/action
http://localhost/ay/xxxx
http://localhost/ay/xxxxx.do
```

路径匹配和扩展名匹配无法同时设置，如果设置，启动 tomcat 服务器会报错。例如下面 3 个匹配规则是错误的：

```
<url-pattern>/kata/*.jsp</url-pattern>
<url-pattern>/*.jsp</url-pattern>
<url-pattern>he*.jsp</url-pattern>
```

（4）默认匹配

```
<servlet-mapping>
    <servlet-name>spring-dispatcher</servlet-name>
    <!-- 默认匹配所有的请求 -->
    <url-pattern>/</url-pattern>
</servlet-mapping>
```

（5）匹配顺序

当一个 URL 与多个 Servlet 的匹配规则可以匹配时，则按照“精确路径 > 最长路径 > 扩展名”这样的优先级匹配到对应的 Servlet。举例如下：

- 比如ServletA的url-pattern为/test，ServletB的url-pattern为/*，如果访问的URL路径为http://localhost/test，容器会优先进行精确路径匹配，发现/test正好被ServletA精确匹配，那么就会调用ServletA，而不是ServletB。
- 比如ServletA的url-pattern为/test/*，而ServletB的url-pattern为/test/a/*，如果访问的URL路径为http://localhost/test/a，容器会选择路径最长的Servlet来匹配，也就是ServletB。
- 比如ServletA的url-pattern：*.action，ServletB的url-pattern为/*。如果访问的URL路径为http://localhost/test.action，容器会优先进行路径匹配，而不是扩展名匹配，这样就去调用ServletB。

接着，我们在/src/main/resources 目录下创建配置文件 spring-mvc.xml，具体代码如下：

```
<?xml version="1.0" encoding="UTF-8"?>
<beans xmlns="http://www.springframework.org/schema/beans"
       xmlns:xsi="http://www.w3.org/2001/XMLSchema-instance"
       xmlns:context="http://www.springframework.org/schema/context"
       xmlns:mvc="http://www.springframework.org/schema/mvc"
       xmlns:aop="http://www.springframework.org/schema/aop"
       xsi:schemaLocation="http://www.springframework.org/schema/beans
        http://www.springframework.org/schema/beans/spring-beans.xsd
        http://www.springframework.org/schema/context
        http://www.springframework.org/schema/context/spring-context.xsd
        http://www.springframework.org/schema/mvc
        http://www.springframework.org/schema/mvc/spring-mvc.xsd
        http://www.springframework.org/schema/aop
        http://www.springframework.org/schema/aop/spring-aop.xsd">
```

```
    <!-- 扫描 controller(后端控制器),并且扫描其中的注解-->
    <context:component-scan base-package="com.ay.controller"/>
    <!--设置配置方案 -->
    <mvc:annotation-driven/>

      <!--配置 JSP  显示 ViewResolver(视图解析器)-->
   <bean class="org.springframework.web.servlet.view.
InternalResourceViewResolver">
       <property name="viewClass"
   value="org.springframework.web.servlet.view.JstlView"/>
       <property name="prefix" value="/WEB-INF/views/"/>
       <property name="suffix" value=".jsp"/>
    </bean>
  </beans>
```

- <context:component-scan>：扫描base-package包或者子包下所有的controller类，并把匹配的controller类注册成Bean。
- <mvc:annotation-driven/>：该注解会自动注册RequestMappingHandlerMapping和 RequestMappingHandlerAdapter两个Bean，是Spring MVC为@Controller分发请求所必须的，并提供了数据绑定支持、@NumberFormatannotation支持、@DateTimeFormat支持、@Valid支持、读写XML的支持（JAXB）和读写JSON的支持（Jackson）等功能。
- InternalResourceViewResolver：最常用的视图解析器，当控制层返回“hello”时，InternalResourceViewResolver解析器会自动添加前缀和后缀，最终路径结果为：/WEB-INF/views/hello.jsp。

最后，在 /src/main/java 目录下创建包 com.ay.controller，并创建控制层类AyTestController，具体代码如下：

```
package com.ay.controller;
import org.springframework.web.bind.annotation.GetMapping;
import org.springframework.web.bind.annotation.RequestMapping;
import org.springframework.web.bind.annotation.RestController;
/**
 *@author Ay
 * @date 2018/04/02
 */
@Controller
@RequestMapping("/test")
public class AyTestController {
   @GetMapping("/sayHello")
   public String sayHello(){
       return "hello";
   }
}
```

- @Controller：标明AyTestController是一个控制器类，使用@Controller标记的类就是一个Spring MVC Controller对象。
- @RequestMapping：是一个用来处理请求地址映射的注解，可用于类或者方法上。用于类上，表示类中的所有响应请求的方法都是以该地址作为父路径。@RequestMapping注解有value、method等属性，value属性可以默认不写。"/test"就是value属性的值。value属性的值就是请求的实际地址。
- @GetMapping：@GetMapping是一个组合注解，是@RequestMapping(method = RequestMethod.GET)的缩写。该注解将HTTP Get 请求映射到特定的处理方法上。类似的@PostMapping注解是@RequestMapping(method = RequestMethod.POST)的缩写。@PutMapping注解是@RequestMapping(method = RequestMethod.PUT)的缩写。@DeleteMapping注解是@RequestMapping(method = RequestMethod.DELETE)的缩写。@PatchMapping 注解是@RequestMapping(method = RequestMethod.PATCH)的缩写。

在/src/main/webapp/WEB-INF 目录下创建 views 文件夹，在 views 文件下创建 hello.jsp 文件，具体代码如下：

```
    <%@page language="java" contentType="text/html; charset=UTF-8"
pageEncoding="UTF-8" %>
    <!DOCTYPE HTML>
    <html>
    <head>
        <title>Getting Started: Serving Web Content</title>
        <meta http-equiv="Content-Type" content="text/html; charset=UTF-8" />
    </head>
    <body>

    hello, ay

    </body>
    </html>
```

至此，Web 项目集成 Spring MVC 大功告成。我们把 Web 项目部署到 Tomcat 服务器上，成功启动 Tomcat 服务器后，在浏览器输入访问路径：http://localhost:8080/test/sayHello。当出现如图 2-7 所示的结果时，代表 Web 项目集成 Spring MVC 成功。

图 2-7　集成 Spring MVC 框架测试

2.4 集成 MyBatis 框架

在 2.3 节中，我们已经在 Web 项目中集成了 Spring MVC，这一节主要介绍如何在 Web 项目中集成 MyBatis 框架。

首先，把集成 MyBatis 框架所需要的依赖包添加到 pom.xml 文件中，具体代码如下：

```
<properties>
   <spring.version>5.0.4.RELEASE</spring.version>
   <javax.servlet.version>4.0.0</javax.servlet.version>
   <jstl.version>1.2</jstl.version>
   <mybatis.version>3.4.6</mybatis.version>
   <mysql.connector.java.version>8.0.9-rc</mysql.connector.java.version>
   <druid.version>1.1.9</druid.version>
   <mybatis.spring.version>1.3.2</mybatis.spring.version>
</properties>
<!--mybatis start -->
<dependency>
     <groupId>mysql</groupId>
     <artifactId>mysql-connector-java</artifactId>
     <version>${mysql.connector.java.version}</version>
     <scope>runtime</scope>
</dependency>

<dependency>
     <groupId>com.alibaba</groupId>
     <artifactId>druid</artifactId>
     <version>${druid.version}</version>
</dependency>

<dependency>
     <groupId>org.springframework</groupId>
     <artifactId>spring-jdbc</artifactId>
     <version>${spring.version}</version>
 </dependency>

 <dependency>
     <groupId>org.mybatis</groupId>
     <artifactId>mybatis</artifactId>
     <version>${mybatis.version}</version>
 </dependency>

<dependency>
```

```
        <groupId>org.mybatis</groupId>
        <artifactId>mybatis-spring</artifactId>
        <version>${mybatis.spring.version}</version>
    </dependency>
  <!--mybatis end -->
```

- mysql-connector-java：是MySQL的JDBC驱动包，用JDBC连接MySQL数据库时必须使用该jar包。
- Druid：Druid是阿里巴巴开源平台上一个数据库连接池实现，它结合了C3P0、DBCP、PROXOOL等DB连接池的优点，同时加入了日志监控，可以很好地监控DB连接池和SQL的执行情况，可以说是针对监控而生的DB连接池，据说是目前最好的连接池。
- mybatis-spring：mybatis-spring会帮助你将MyBatis代码无缝地整合到Spring中，使用这个类库中的类，Spring将会加载必要的MyBatis工厂类和Session类。

其次，在/src/main/resources目录下创建jdbc.properties配置文件，具体代码如下：

```
    //驱动
    jdbc.driverClassName=com.mysql.jdbc.Driver
    //MySQL 连接信息
    jdbc.url=jdbc:mysql://127.0.0.1:3306/springmvc-mybatis-
book?serverTimezone=GMT
    //用户名
    jdbc.username=root
    //密码
    jdbc.password=123456
```

关于jdbc.properties配置主要指配置驱动和连接数据库的配置信息。

最后，在applicationContext.xml配置文件添加如下的配置，具体代码如下：

```
    <!--1.配置数据库相关参数-->
    <context:property-placeholder
    location="classpath:jdbc.properties" ignore-unresolvable="true"/>

    <!--2.数据源 druid -->
    <bean id="dataSource" class="com.alibaba.druid.pool.DruidDataSource"
    init-method="init" destroy-method="close">
        <property name="driverClassName" value="${jdbc.driverClassName}" />
        <property name="url" value="${jdbc.url}" />
        <property name="username" value="${jdbc.username}" />
        <property name="password" value="${jdbc.password}" />
     </bean>
    <!--3.配置 SqlSessionFactory 对象-->
    <bean id="sqlSessionFactory" class="org.mybatis.spring.
SqlSessionFactoryBean">
        <!--注入数据库连接池-->
```

```
        <property name="dataSource" ref="dataSource"/>
        <!--扫描 sql 配置文件:mapper 需要的 xml 文件-->
        <property name="mapperLocations" value="classpath:mapper/*.xml"/>
    </bean>

    <bean id="sqlSession" class="org.mybatis.spring.SqlSessionTemplate">
        <constructor-arg index="0" ref="sqlSessionFactory" />
    </bean>

    <!-- 扫描 basePackage 下所有以@MyBatisDao 注解的接口 -->
    <bean id="mapperScannerConfigurer"
                class="org.mybatis.spring.mapper. MapperScannerConfigurer">
        <property name="sqlSessionFactoryBeanName" value="sqlSessionFactory" />
        <property name="basePackage" value="com.ay.dao"/>
    </bean>
```

- <context:property-placeholder/>：该标签提供了一种优雅的外在化参数配置的方式，location表示属性文件位置，多个属性文件之间通过逗号分隔。ignore-unresolvable表示是否忽略解析不到的属性，如果不忽略，找不到将抛出异常。
- DruidDataSource：阿里巴巴Druid数据源，该数据源会读取jdbc.properties配置文件的数据库连接信息和驱动。
- SqlSessionFactoryBean：在基本的MyBatis中，Session工厂可以使用SqlSessionFactoryBuilder来创建；而在MyBatis-Spring中，则使用SqlSessionFactoryBean来替代。

SqlSessionFactoryBean 需要依赖数据源 dataSource。mapperLocations 属性可以用来指定 MyBatis 的 XML 映射器文件的位置，值为 mapper/*.xml 代表扫描 classpath 路径下 mapper 文件夹下的所有 XML 文件。

- MapperScannerConfigurer：没有必要在Spring的XML配置文件中注册所有的映射器。相反，可以使用MapperScannerConfigurer，它将会查找类路径下的映射器并自动将它们创建成MapperFactoryBean。basePackage属性是让你为映射器接口文件设置基本的包路径。可以使用分号或逗号作为分隔符设置多于一个的包路径。每个映射器将会在指定的包路径中递归地被搜索到。这里设置的值是com.ay.dao这个包。

Web 应用集成 MyBatis 框架所需的配置文件都添加完成之后，我们开始开发相关的代码。

首先，在 MySQL 数据库创建表 ay_user，具体的 SQL 语句如下：

```
- ----------------------------
-- Table structure for ay_user
-- ----------------------------
DROP TABLE IF EXISTS 'ay_user';
```

```
CREATE TABLE 'ay_user' (
  'id' bigint(32) NOT NULL AUTO_INCREMENT,
  'name' varchar(10) DEFAULT NULL,
  'password' varchar(64) DEFAULT NULL,
  PRIMARY KEY ('id')
) ENGINE=InnoDB AUTO_INCREMENT=2 DEFAULT CHARSET=utf8;
```

数据库表创建完成之后，往 ay_user 表插入数据，具体如图 2-8 所示。

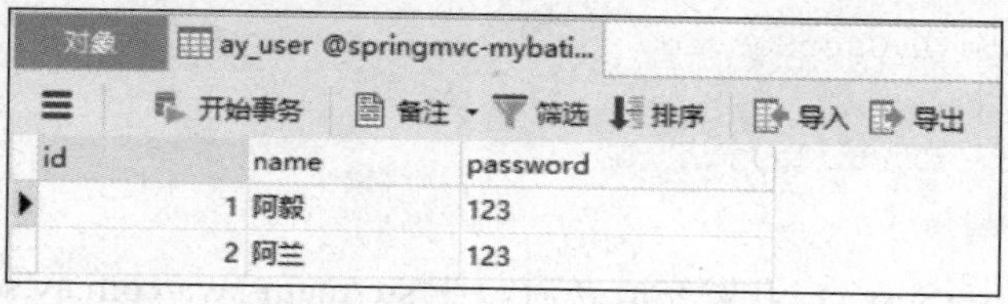

对象 | ay_user @springmvc-mybati...

开始事务 备注 筛选 排序 导入 导出

id	name	password
1	阿毅	123
2	阿兰	123

图 2-8 用户数据

数据库表创建完成之后，在/src/main/java/com.ay.model 目录下创建数据库表对应的实体类对象 AyUser，具体的代码如下：

```
/**
 * 用户实体
 * @author Ay
 * @date  2018/04/02
 */
public class AyUser implements Serializable{

    private Integer id;
    private String name;
    private String password;

    //省略 set、get 方法
}
```

实体类对象 AyUser 创建完成之后，在/src/main/java/ com.ay.dao 目录下创建对应的 DAO 对象 AyUserDao，AyUserDao 是一个接口，提供了 findAll 方法用来查询所有的用户。AyUserDao 具体代码如下：

```
package com.ay.dao;
import com.ay.model.AyUser;
import org.springframework.stereotype.Repository;
import java.util.List;

@Repository
public interface AyUserDao {

    List<AyUser> findAll();
}
```

接口类AyUserDao创建完成之后，在/src/main/java/com.ay.service目录下创建对应的服务层接口AyUserService，服务层接口AyUserService代码也非常简单，只提供了一个查询所有用户的方法findAll()，具体的代码如下：

```
package com.ay.service;
import com.ay.model.AyUser;
import java.util.List;

public interface AyUserService {

    List<AyUser> findAll();
}
```

服务层接口AyUserService开发完成之后，在/src/main/java/com.ay.service.impl开发对应的服务层实现类AyUserServiceImpl，实现类主要是注入AyUserDao接口，并实现findAll()方法，在findAll()方法中调用AyUserDao的findAll()方法，具体代码如下：

```
package com.ay.service.impl;
import com.ay.dao.AyUserDao;
import com.ay.model.AyUser;
import com.ay.service.AyUserService;
import org.springframework.stereotype.Repository;
import org.springframework.stereotype.Service;
import javax.annotation.Resource;
import java.util.List;
@Service
public class AyUserServiceImpl implements AyUserService{

    @Resource
    private AyUserDao ayUserDao;

    public List<AyUser> findAll() {
        return ayUserDao.findAll();
    }
}
```

服务层实现类AyUserServiceImpl开发完成之后，在/src/main/java/com.ay.controller目录下创建控制层类AyUserController，并注入服务层接口。AyUserController类只有一个findAll()方法。在AyUserController类上添加映射路径/user，在findAll()方法上添加映射路径/findAll。

```
package com.ay.controller;
import com.ay.model.AyUser;
import com.ay.service.AyUserService;
import org.springframework.stereotype.Controller;
import org.springframework.ui.Model;
```

```
import org.springframework.web.bind.annotation.GetMapping;
import org.springframework.web.bind.annotation.RequestMapping;
import javax.annotation.Resource;
import java.util.List;

/**
 *@author Ay
 * @date 2018/04/02
 */
@Controller
@RequestMapping(value = "/user")
public class AyUserController {

    @Resource
    private AyUserService ayUserService;

    @GetMapping("/findAll")
    public String findAll(Model model){
        List<AyUser> ayUserList = ayUserService.findAll();
          for(AyUser ayUser : ayUserList){
            System.out.println("id: " + ayUser.getId());
            System.out.println("name: " + ayUser.getName());
        }
        return "hello";
    }
}
```

最后，在/src/main/resources/mapper 目录下创建 AyUserMapper.xml 文件，具体代码如下：

```
<?xml version="1.0" encoding="UTF-8" ?>
<!DOCTYPE mapper PUBLIC "-//mybatis.org//DTD Mapper 3.0//EN"
        "http://mybatis.org/dtd/mybatis-3-mapper.dtd">
<mapper namespace="com.ay.dao.AyUserDao">
    <sql id="userField">
        a.id as "id",
        a.name as "name",
        a.password as "password"
    </sql>
    <!-- 获取所有用户 -->
    <select id="findAll" resultType="com.ay.model.AyUser">
        select
        <include refid="userField"/>
        from ay_user as a
    </select>
```

```
</mapper>
```

- <mapper/>：namespace主要用于绑定Dao接口，这里绑定com.ay.dao.AyUserDao接口。
- <select id="findAll">：select标签，用来编写select查询语句，id属性值与AyUserDao接口中的方法名一一对应。在select标签中，查询了ay_user表中的所有数据，并返回。

到这里，Web 应用集成 Spring、Spring MVC、MyBatis 已经全部完成，现在重新启动 Tomcat 服务器，在浏览器输入访问地址：http://localhost:8080/user/findAll，如果能看到如图 2-9 和图 2-10 所示的信息，代表整合成功。

图 2-9　浏览器输出信息

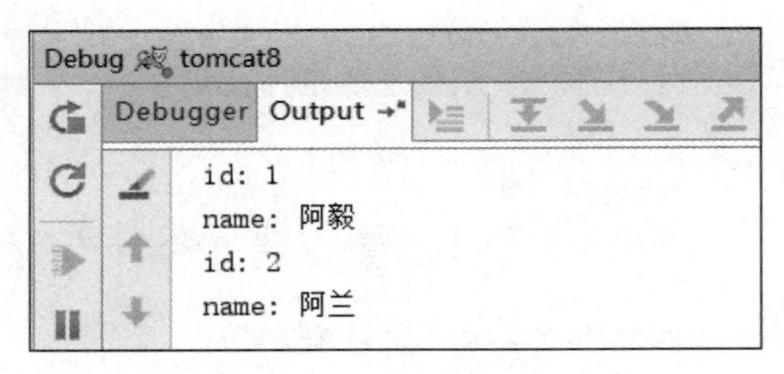

图 2-10　控制台打印信息

2.5　集成 Log4j 日志框架

Log4j 是 Apache 下的一个开源项目，通过使用 Log4j 可以将日志信息打印到控制台、文件等。也可以控制每一条日志的输出格式，通过定义每一条日志信息的级别，能够更加细致地控制日志的生成过程。

在应用程序中添加日志记录有三个目的：

（1）监视代码中变量的变化情况，周期性地记录到文件中供其他应用进行统计分析工作。

（2）跟踪代码运行时轨迹，作为日后审计的依据。

（3）担当集成开发环境中的调试器的作用，向文件或控制台打印代码的调试信息。

Log4j 中有三个主要的组件，它们分别是：Logger（记录器）、Appender（输出端）和 Layout（布局），这三个组件可以简单地理解为日志类别，日志要输出的地方和日志以何种形式输出。Log4j 原理如图 2-11 所示。

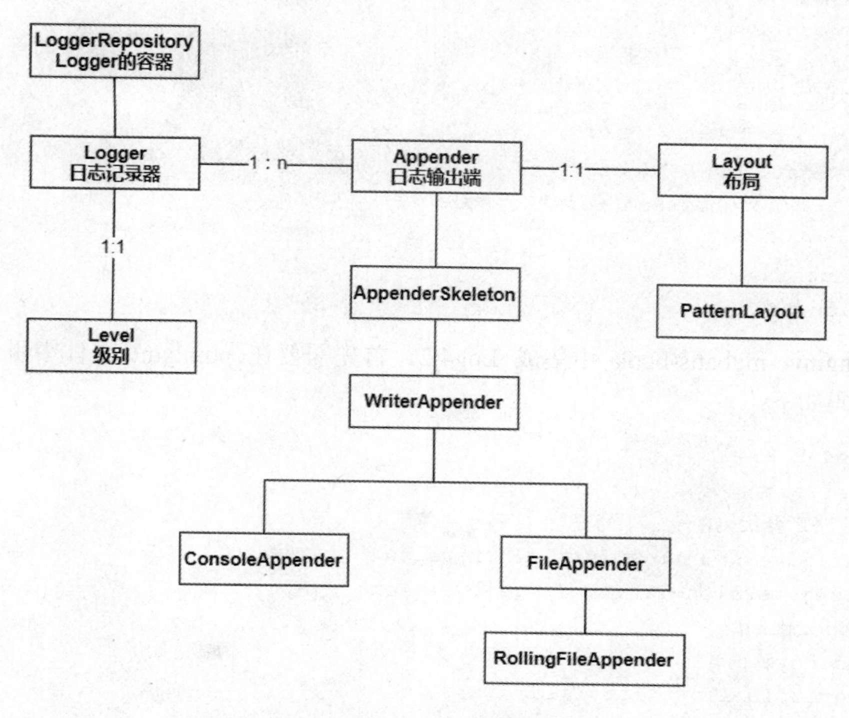

图 2-11　Log4j 日志框架简单原理图

- **Logger（记录器）**：Logger组件被分为7个级别：all、debug、info、warn、error、fatal、off。这7个级别是有优先级的：all<debug< info< warn< error< fatal<off，分别用来指定这条日志信息的重要程度。Log4j有一个规则：只输出级别不低于设定级别的日志信息。假设Logger级别设定为info，则info、warn、error和fatal级别的日志信息都会输出，而级别比info低的debug则不会输出。Log4j允许开发人员定义多个Logger，每个Logger拥有自己的名字，Logger之间通过名字来表明隶属关系。
- **Appender（输出端）**：Log4j日志系统允许把日志输出到不同的地方，如控制台（Console）、文件（Files）等，可以根据天数或者文件大小产生新的文件，可以以流的形式发送到其他地方等。
- **Layout（布局）**：Layout的作用是控制Log信息的输出方式，也就是格式化输出的信息。

Log4j 支持两种配置文件格式，一种是 XML 格式的文件；另一种是 Java 特性文件 log4j2.properties（键 = 值），properties 文件简单易读，而 XML 文件可以配置更多的功能（比如过滤），没有好坏，能够融会贯通就是最好的。具体的 XML 配置如下：

```
<?xml version="1.0" encoding="UTF-8"?>
<Configuration status="WARN">
    <Appenders>
        <Console name="Console" target="SYSTEM_OUT">
            <PatternLayout pattern="%d{HH:mm:ss.SSS} [%t] %-5level
```

```
%logger{36} - %msg%n" />
            </Console>
        </Appenders>
        <Loggers>
            <Root level="debug">
                <AppenderRef ref="Console" />
            </Root>
        </Loggers>
    </Configuration>
```

在 springmvc-mybatis-book 中集成 Log4j2，首先需要在 pom.xml 文件中引入所需的依赖，具体代码如下：

```
    !-- log4j2 -->
    <properties>
        //省略部分代码
        <slf4j.version>1.7.7</slf4j.version>
        <log4j.version>1.2.17</log4j.version>
      </properties>
    <dependency>
          <groupId>log4j</groupId>
          <artifactId>log4j</artifactId>
          <version>${log4j.version}</version>
        </dependency>
        <dependency>
          <groupId>org.slf4j</groupId>
          <artifactId>slf4j-api</artifactId>
          <version>${slf4j.version}</version>
        </dependency>
        <dependency>
          <groupId>org.slf4j</groupId>
          <artifactId>slf4j-log4j12</artifactId>
          <version>${slf4j.version}</version>
        </dependency>
```

- slf4j-api：全称Simple Logging Facade For Java，为Java提供的简单日志Facade。Façade门面就是接口，它允许用户以自己的喜好在项目中通过slf4j接入不同的日志系统。更直观一点，slf4j是个数据线，一端嵌入程序，另一端链接日志系统，从而实现将程序中的信息导入到日志系统并记录。
- slf4j-log4j12：链接slf4j-api和log4j的适配器。
- log4j：具体的日志系统。通过slf4j-log4j12初始化log4j，达到最终日志的输出。

slf4j-api、slf4j-log4j12 和 log4j 之间的关系如图 2-12 所示。

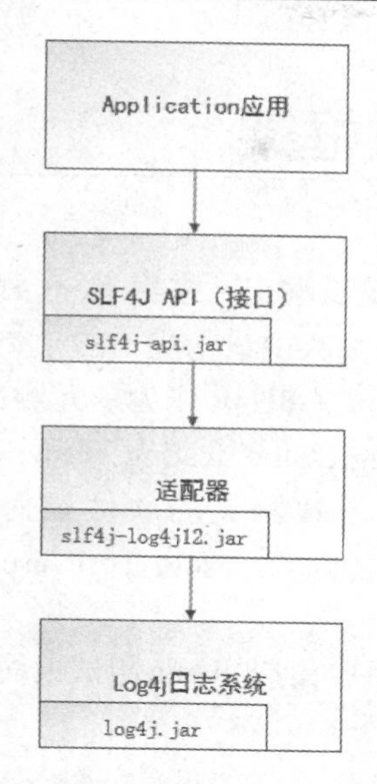

图 2-12　slf4j-api、slf4j-log4j12 和 log4j 的关系描述

集成 Log4j 的依赖包添加完成之后，在项目的/src/main/java/resources/下创建配置文件 log4j.properties，具体代码如下：

```
###set log levels
log4j.rootLogger = DEBUG,Console

###输出到控制台
log4j.appender.Console=org.apache.log4j.ConsoleAppender
log4j.appender.Console.Target=System.out
log4j.appender.Console.layout=org.apache.log4j.PatternLayout
log4j.appender.Console.layout.ConversionPattern= %d{ABSOLUTE} %5p %c{1}:%L
- %m%n
```

在 log4j.properties 配置文件中，设置了日志级别和将日志输出到控制台。现在重新启动 springmvc-mybatis-book 项目，当在控制台看到相关的日志打印信息时，表示成功集成了 Log4j 日志框架，具体如图 2-13 所示。

```
08:32:16,703 DEBUG DefaultListableBeanFactory:219 - Creating shared instance of singleto
08:32:16,705 DEBUG DefaultListableBeanFactory:467 - Creating instance of bean 'ayUserDao
08:32:16,705 DEBUG DefaultListableBeanFactory:573 - Eagerly caching bean 'ayUserDao' to
08:32:16,706 DEBUG DefaultListableBeanFactory:255 - Returning cached instance of singlet
08:32:16,707 DEBUG DefaultListableBeanFactory:1755 - Invoking afterPropertiesSet() on be
08:32:16,708 DEBUG DefaultListableBeanFactory:504 - Finished creating instance of bean '
08:32:16,712 DEBUG DefaultListableBeanFactory:504 - Finished creating instance of bean '
08:32:16,712 DEBUG DefaultListableBeanFactory:504 - Finished creating instance of bean '
08:32:16,713 DEBUG DefaultListableBeanFactory:219 - Creating shared instance of singleto
```

图 2-13　控制台打印的日志信息

2.6 集成 JUnit 测试框架

JUnit 是一个 Java 语言的单元测试框架。它由 Kent Beck 和 Erich Gamma 建立，逐渐成为源于 Kent Beck 的 sUnit 的 xUnit 家族中最为成功的一个。JUnit 有它自己的 JUnit 扩展生态圈，多数 Java 的开发环境都已经集成了 JUnit 作为单元测试的工具。

JUnit 是一个回归测试框架（regression testing framework）。JUnit 测试是程序员测试，即所谓白盒测试，因为程序员知道被测试的软件如何（How）完成功能和完成什么样（What）的功能。由于 JUnit 是一套框架，继承 TestCase 类，所以可以用 JUnit 进行自动测试。

在 springmvc-mybatis-book 项目中集成 JUnit 测试很简单，首先在项目的 pom.xml 配置文件中添加相关的依赖，具体代码如下：

```
<!-- junit -->
<dependency>
    <groupId>junit</groupId>
    <artifactId>junit</artifactId>
    <version>4.12</version>
</dependency>
```

然后，在项目的/src/main/test/com.ay.test 目录下创建测试基类 BaseJunit4Test，具体代码如下：

```
/**
 * 描述：测试基类
 * @author Ay
 * @create 2018/05/04
 **/
@RunWith(SpringJUnit4ClassRunner.class)
@ContextConfiguration(locations={"classpath:applicationContext.xml"})
public class BaseJunit4Test {
}
```

- @RunWith：参数化运行器，用于指定JUnit运行环境，是JUnit提供给其他框架测试环境的接口扩展，为了便于使用Spring的依赖注入，Spring提供了SpringJUnit4ClassRunner作为JUnit测试环境。
- @ContextConfiguration：加载配置文件applicationContext.xml。

BaseJunit4Test 类开发完成之后，在/src/main/test/com.ay.test 目录下创建 AyUserDaoTest 测试类，简单测试集成 JUnit 框架是否成功，具体代码如下：

```
/**
 * 描述：用户 DAO 测试类
 * @author Ay
```

```
 * @create 2018/05/04
 **/
public class AyUserDaoTest extends BaseJunit4Test{

    @Resource
    private AyUserDao ayUserDao;

    @Test
    public void testFindAll(){
       List<AyUser> userList = ayUserDao.findAll();
       System.out.println(userList.size());
    }
}
```

AyUserDaoTest 类需要继承 BaseJunit4Test 测试基类。AyUserDaoTest 类代码开发完成之后，运行 testFindAll 方法，便可以在控制台看到相关的打印信息。

2.7　思考与练习

1. 请动手搭建一个 SSM 项目。
2. 简述 applicationContext.xml 配置文件中的<context:component-scan/>标签的作用。
3. 简述 web.xml 配置文件中的<context-param>标签的作用。
4. 简述 ContextLoaderListener 类的具体作用。
5. ApplicationContext 接口有哪些实现类？
6. 简述 DispatcherServlet 类的具体作用。
7. 简单描述 Log4j 日志框架的原理。
8. 简述应用框架记录日志的目的。
9. Log4j 支持哪些配置文件的格式？
10. 简述 JUnit 测试框架的内容。
11. 动手开发一个测试类，并打印简单的字符串。

第 3 章

Spring 核心 IOC 与 AOP

本章主要回顾 Spring 的基础知识 IOC 和 AOP、IOC 和 AOP 背后的实现原理和设计模式。这些设计模式包括单例模式、简单工厂模式、工厂方法模式和动态代理模式等。

3.1 Spring IOC 和 DI

3.1.1 Spring IOC 和 DI 概述

学习 Spring，经常会联系到 Spring 的 IOC（控制反转）和 DI（依赖注入）。在 Spring 环境下这两个概念是等同的，控制反转是通过依赖注入来实现的。

IOC 是指原先我们代码里面需要实现的对象创建、维护对象间的依赖关系，反转给容器来帮忙实现。那么必然的我们需要创建一个容器，同时需要一种描述来让容器知道需要创建的对象与对象的关系。依赖注入的目的是为了解耦，体现一种“组合”的理念。继承一个父类，子类将与父类耦合，组合关系使耦合度大大降低。

Spring IOC 容器负责创建 Bean，并通过容器将 Bean 注入到需要的 Bean 对象上。同时 Spring IOC 容器还负责维护 Bean 对象之间的关系。那么，Spring IOC 如何来体现对象与对象之间的关系呢？Spring 提供了 XML 配置和 Java 配置等方式，具体看下面的实例：

```
@Service
public class AyUserServiceImpl implements AyUserService{

    @Resource
    private AyUserDao ayUserDao;
```

```
    public List<AyUser> findAll() {
        return ayUserDao.findAll();
    }
}
```

Spring 提供的注解有很多，比如声明 Bean 的注解和注入 Bean 的注解，这些注解在工作中经常被使用，所以有必要在这里重新回顾一下。

声明 Bean 的注解：

- @Component：没有明确的角色。
- @Service：在服务层（业务逻辑层）被使用。
- @Repository：在数据访问层（dao层）被使用。
- @Controller：在表现层（控制层）被使用。

注入 Bean 的注解：

- @Autowired：Spring提供的注解。
- @Resource：JSR-250提供的注解。

注意，@Resource 这个注解属于 J2EE 的，默认按照名称进行装配，名称可以通过 name 属性进行指定。如果没有指定 name 属性，当注解写在字段上时，默认取字段名进行查找。如果注解写在 setter 方法上默认取属性名进行装配。当找不到与名称匹配的 Bean 时才按照类型进行装配。但是需要注意的是，如果 name 属性一旦指定，就只会按照名称进行装配。具体代码如下：

```
@Resource(name = "ayUserDao")
private AyUserDao ayUserDao;
```

而@Autowired 这个注解是属于 Spring 的，默认按类型装配。默认情况下要求依赖对象必须存在，如果要允许 null 值，可以设置它的 required 属性为 false，如@Autowired(required=false)，如果我们想使用名称装配，可以结合@Qualifier 注解进行使用。具体代码如下：

```
@Autowired
@Qualifier("ayUserDao")
private AyUserDao ayUserDao;
```

3.1.2 单例模式

Spring 依赖注入 Bean 实例默认都是单例的，所以有必要来回顾一下单例模式。

对于一个软件系统的某些类而言，无须创建多个实例，例如 Windows 任务管理器，如图 3-1 所示。

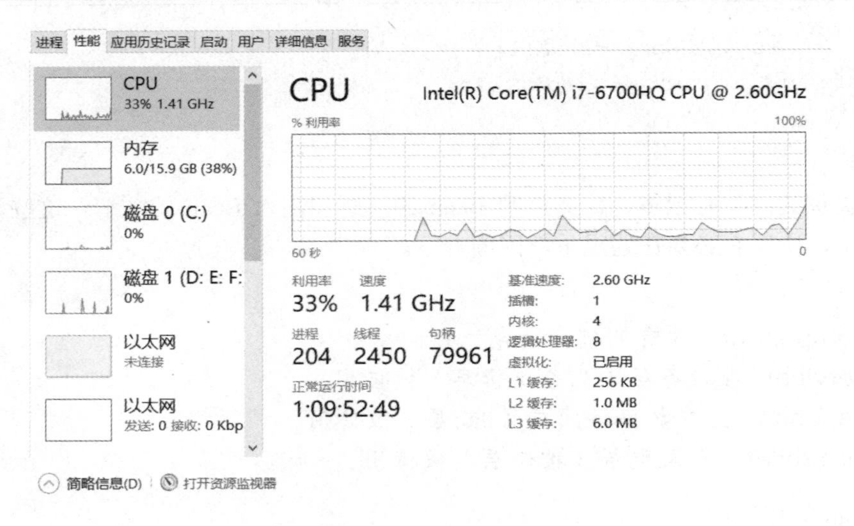

图 3-1　Windows 任务管理器

我们没办法打开多个任务管理器，也就是说，在一个 Windows 系统中，任务管理器存在唯一性。为什么要这样设计呢？可以从以下两个方面来分析：其一，如果能弹出多个窗口，且这些窗口的内容完全一致，全部是重复对象，势必会浪费系统资源，任务管理器需要获取系统运行时的诸多信息，这些信息的获取需要消耗一定的系统资源，包括 CPU 资源及内存资源等，浪费是可耻的，而且根本没有必要显示多个内容完全相同的窗口；其二，如果弹出的多个窗口内容不一致，问题就更加严重了，这意味着在某一瞬间系统资源的使用情况和进程、服务等信息存在多个状态，例如任务管理器窗口 A 显示“CPU 使用率”为 10%，窗口 B 显示“CPU 使用率”为 15%，到底哪个才是真实的呢？这纯属“调戏”用户，给用户带来误解，很不可取。由此可见，确保 Windows 任务管理器在系统中有且仅有一个非常重要。除了任务管理器外，数据库连接池、应用配置等都是使用单例的。

了解完单例模式的使用场景后，再来看看单例模式的定义：单例模式（Singleton Pattern）确保某一个类只有一个实例，而且自行实例化并向整个系统提供这个实例，这个类称为单例类，它提供全局访问的方法。单例模式是一种对象创建型模式。

先来看一个传统的创建类的代码：

```
/**
 * 描述：传统创建类实例
 * @author Ay
 * @create 2018/1/23
 */
public class Case_1 {

    public static void main(String[] args) {
        Singleton singleton = new Singleton();
Singleton singleton2 = new Singleton();
    }
}
```

```
/**
 * 描述：单例类
 */
class Singleton{

}
```

上述代码中，每次 new Singleton()都会创建一个 Singleton 实例，显然不符合一个类只有一个实例的要求，所以需要对上述代码进行修改，具体修改如下：

```
/**
 * 描述：单例模式实例
 * @author Ay
 * @create 2018/1/23
 */
public class Case_1 {
    public static void main(String[] args) {
        //Singleton singleton = new Singleton();
        //单例
        Singleton singleton = Singleton.getInstance();
    }
}

/**
 * 描述：单例类（饿汉模式）
 */
class Singleton{
    //step 2. 自行对外提供实例
    private static final Singleton singleton = new Singleton();
    //step 1. 构造函数私有化
    private Singleton(){}
    //step 3. 提供外界可以获得该实例的方法
    public static Singleton getInstance(){
        return singleton;
    }
}
```

单例模式的写法有很多种，上述代码是一个最简单的饿汉模式的实现方法，在类加载的时候就创建了单例类的对象。由上述代码可知，实现一个单例模式总共有三步：

（1）构造函数私有化。
（2）自行对外提供实例。
（3）提供外界可以获得该实例的方法。

与饿汉模式相对应的还有懒汉模式，懒汉模式有延迟加载的意思，具体代码如下：

```
/**
```

```
 * 描述：懒汉模式（存在多线程并发的问题，不是正确的写法）
 * @author Ay
 * @create 2018/04/14
 */
class Singleton{
    private static Singleton singleton = null;
    private Singleton(){}
    public static Singleton getInstance(){
        //1. 判断对象是否创建
        if(null == singleton){
            //2. 创建对象
            singleton = new Singleton();
        }
        return singleton;
    }
}
```

如果创建单例对象会消耗大量资源，那么延迟创建对象是一个不错的选择，但是懒汉模式有一个明显的问题，就是没有考虑线程安全问题，在多线程并发的情况下，会并发调用 getInstance()方法，从而导致系统同时创建多个单例类实例，显然不符合要求。可以通过给 getInstance()方法添加锁解决该问题，具体代码如下：

```
/**
 * 描述：懒汉模式（添加 synchronized 锁）
 * @author Ay
 * @create 2018/04/14
 */
class Singleton{
    private static Singleton singleton = null;
    private Singleton(){}
    public static synchronized Singleton getInstance(){
        //1. 判断对象是否创建
        if(null == singleton){
            //2. 创建对象
            singleton = new Singleton();
        }
        return singleton;
    }
}
```

添加 synchronized 锁虽然可以保证线程安全，但是每次访问 getInstance()方法的时候，都会有加锁和解锁操作，同时 synchronized 锁是添加在方法上面，锁的范围过大，而单例类是全局唯一的，锁的操作会成为系统的瓶颈。因此，需要对代码再进行优化，由此引出了“双重校验锁”的方式，具体代码如下：

```
/**
```

```
 * 描述：双重校验锁（指令重排问题）
 * @author Ay
 * @create 2018/04/14
 */
class Singleton{
    private static Singleton singleton = null;
    private Singleton(){}
    public static  Singleton getInstance(){
        //第一次校验
        if(singleton == null){
            synchronized(Singleton.class){
                //第二次检验
                if(singleton == null){
                    //创建对象，非原子操作
                    singleton = new Singleton();
                }
            }
        }
        return singleton;
    }
}
```

双重校验锁会出现指令重排的问题，所谓指令重排是指：JVM 为了优化指令，提高程序运行效率，在不影响单线程程序执行结果的前提下，尽可能地提高并行度。singleton = new Singleton()看似原子操作，其实不然，singleton = new Singleton()实际上可以抽象为下面几条 JVM 指令：

```
//1：分配对象的内存空间
memory = allocate();
//2：初始化对象
ctorInstance(memory);
//3：设置 instance 指向刚分配的内存地址
singleton = memory;
```

上面操作 2 依赖于操作 1，但是操作 3 并不依赖于操作 2，所以 JVM 是可以针对它们进行指令的优化重排序的，经过重排序后如下：

```
//1：分配对象的内存空间
memory = allocate();
//3：instance 指向刚分配的内存地址，此时对象还未初始化
singleton = memory;
//2：初始化对象
ctorInstance(memory);
```

可以看到，指令重排之后，singleton 指向分配好的内存放在了前面，而这段内存的初始化被排在了后面。在线程 A 执行这段赋值语句，在初始化分配对象之前就已经将其赋值给

singleton 引用，恰好 B 线程进入方法判断 singleton 引用不为 null，然后就将其返回使用，导致程序出错。

为了解决指令重排的问题，可以使用 volatile 关键字修饰 singleton 字段。volatile 关键字的一个语义就是禁止指令的重排序优化，阻止 JVM 对其相关代码进行指令重排，这样就能够按照既定的顺序指令执行。修改后的代码如下：

```
/**
 * 描述：双重校验锁（volatile 解决指令重排问题）
 * @author Ay
 * @create 2018/04/14
 */
class Singleton{
    private static volatile  Singleton singleton = null;
    private Singleton(){}
    public static  Singleton getInstance(){
        if(singleton == null){
            synchronized(Singleton.class){
                if(singleton == null){
                    singleton = new Singleton();
                }
            }
        }
      return singleton;
    }
}
```

除了双重校验锁的写法外，比较推荐大家使用最后一种单例模式的写法：静态内部类的单例模式，具体代码如下：

```
/**
 * 描述：静态内部类单例模式（推荐的写法）
 * @author Ay
 * @create 2018/04/14
 */
class ___Singleton{

    //2：私有的静态内部类，类加载器负责加锁
    private static class SingletonHolder{
        private static ___Singleton singleton = new ___Singleton();
    }
    //1：私有化构造方法
    private ___Singleton(){}
    //3：自行对外提供实例
    public static ___Singleton getInstance(){
        return SingletonHolder.singleton;
```

```
        }
    }
```

当第一次访问类中的静态字段时，会触发类加载，并且同一个类只加载一次。静态内部类也是如此，类加载过程由类加载器负责加锁，从而保证线程安全。这种写法相对于双重检验锁的写法，更加简洁明了，也更加不会出错。

3.1.3 Spring 单例模式源码解析

回顾完单例模式的内容，来看一下 Spring 的 BeanFactory 工厂如何实现单例模式。Spring 的依赖注入（包括 lazy-init 方式）都是发生在 AbstractBeanFactory 的 getBean 方法里。getBean 方法内部调用 doGetBean 方法，doGetBean 方法调用 getSingleton 方法进行 bean 的创建。非 lazy-init 方式，在容器初始化时进行调用，lazy-init 方式，在用户向容器第一次索要 bean 时进行调用。AbstractBeanFactory 类源码具体如下：

```
    //省略代码
    @Override
        public Object getBean(String name) throws BeansException {
            return doGetBean(name, null, null, false);
        }

    protected <T> T doGetBean(final String name, @Nullable final Class<T>
requiredType,
            @Nullable final Object[] args, boolean typeCheckOnly) throws
BeansException {

            final String beanName = transformedBeanName(name);
            Object bean;

            //检查单例缓存中是否有手动注册的单例
            Object sharedInstance = getSingleton(beanName);
        //省略代码
    }
    @Nullable
        protected Object getSingleton(String beanName, boolean
allowEarlyReference) {
            Object singletonObject = this.singletonObjects.get(beanName);
            if (singletonObject == null && isSingletonCurrentlyInCreation
(beanName)) {
                synchronized (this.singletonObjects) {
                    singletonObject = this.earlySingletonObjects.get(beanName);
                    if (singletonObject == null && allowEarlyReference) {
                        ObjectFactory<?> singletonFactory
                        = this.singletonFactories.get(beanName);
                        if (singletonFactory != null) {
```

```
                        singletonObject = singletonFactory.getObject();
                        this.earlySingletonObjects.put(beanName,
singletonObject);
                        this.singletonFactories.remove(beanName);
                    }
                }
            }
        }
        return singletonObject;
    }
```

从上面代码中可以看到，Spring 依赖注入时，使用了双重校验锁的单例模式。首先从缓存 singletonObjects（实际上是一个 ConcurrentHashMap）中获取 bean 实例，如果为 null，对缓存 singletonObjects 加锁，然后再从缓存 earlySingletonObjects（实际上是一个 HashMap）中获取 bean 实例，如果继续为 null，就创建一个 bean。在这里 Spring 并没有使用私有构造方法来创建 bean，而是通过 singletonFactory.getObject() 返回具体 beanName 对应的 ObjectFactory 来创建 bean。一路跟踪下去，发现实际上是调用了 AbstractAutowireCapableBeanFactory 的 doCreateBean 方法，返回了 BeanWrapper 包装并创建的 bean 实例。AbstractAutowireCapableBeanFactory 的 doCreateBean 方法部分代码如下：

```
    protected Object doCreateBean(final String beanName,
            final RootBeanDefinition mbd,final @Nullable Object[] args)
            throws BeanCreationException {
        //实例化 bean
        BeanWrapper instanceWrapper = null;
        if (mbd.isSingleton()) {
            instanceWrapper = this.factoryBeanInstanceCache.remove(beanName);
        }
        if (instanceWrapper == null) {
         //创建 bean 实例，并返回 instanceWrapper
            instanceWrapper = createBeanInstance(beanName, mbd, args);
        }
        final Object bean = instanceWrapper.getWrappedInstance();
        Class<?> beanType = instanceWrapper.getWrappedClass();
        if (beanType != NullBean.class) {
            mbd.resolvedTargetType = beanType;
        }
        //允许后处理器修改合并的 bean 定义
        synchronized (mbd.postProcessingLock) {
            if (!mbd.postProcessed) {
                try {
                    applyMergedBeanDefinitionPostProcessors(mbd, beanType,
beanName);
                }
                catch (Throwable ex) {
```

```
                    throw new BeanCreationException
(mbd.getResourceDescription(),
                        beanName,"Post-processing of merged bean definition
failed", ex);
                    }
                    mbd.postProcessed = true;
                }
            }

        //缓存单例能够被解析循环引用，即使被诸如BeanFactoryAware之类的生命周期接口触发

            boolean earlySingletonExposure = (mbd.isSingleton() &&
    this.allowCircularReferences &&
                    isSingletonCurrentlyInCreation(beanName));
            if (earlySingletonExposure) {
                if (logger.isDebugEnabled()) {
                    logger.debug("Eagerly caching bean '" + beanName +
                            "' to allow for resolving potential circular
references");
                }
            //增加beanName和ObjectFactory的键值对应关系
            //获取bean的所有后处理器，并进行处理
            addSingletonFactory(beanName, () -> getEarlyBeanReference(beanName,
mbd, bean));
            }

    //省略大量代码
    }
```

- createBeanInstance(beanName, mbd, args)：创建bean实例并返回instanceWrapper。
- addSingletonFactory：增加beanName和ObjectFactory的键值对应关系。
- getEarlyBeanReference(beanName, mbd, bean)：获取bean的所有后处理器，并进行处理。如果是SmartInstantiationAwareBeanPostProcessor类型，就进行处理，如果没有相关处理内容，就返回默认的实现。

getEarlyBeanReference的具体代码如下：

```
    protected Object getEarlyBeanReference(String beanName, RootBeanDefinition
mbd, Object bean) {
            Object exposedObject = bean;
            if (!mbd.isSynthetic() && hasInstantiationAwareBeanPostProcessors()){
                for (BeanPostProcessor bp : getBeanPostProcessors()) {
                    if (bp instanceof SmartInstantiationAwareBeanPostProcessor){
                        SmartInstantiationAwareBeanPostProcessor ibp
                        = (SmartInstantiationAwareBeanPostProcessor) bp;
                        exposedObject
```

```
                    = ibp.getEarlyBeanReference(exposedObject, beanName);
                }
            }
        }
        return exposedObject;
    }
```

3.1.4 简单工厂模式详解

Spring 的 Bean 工厂大量使用工厂方法模式，而工厂方法模式是以简单工厂模式为基础的，所以有必要先来回顾一下简单工厂模式的基础知识。

简单工厂模式（Simple Factory Pattern）用来定义一个工厂类，它可以根据参数的不同返回不同类的实例，被创建的实例通常都具有共同的父类。因为在简单工厂模式中用于创建实例的方法是静态（static）方法，因此简单工厂模式又被称为静态工厂方法（Static Factory Method）模式，它属于类创建型模式。

简单工厂模式的要点在于，当你需要什么，只需要传入一个正确的参数，就可以获取你所需要的对象，而无须知道其创建细节。简单工厂模式结构比较简单，其核心是工厂类的设计，其结构如图 3-2 所示。

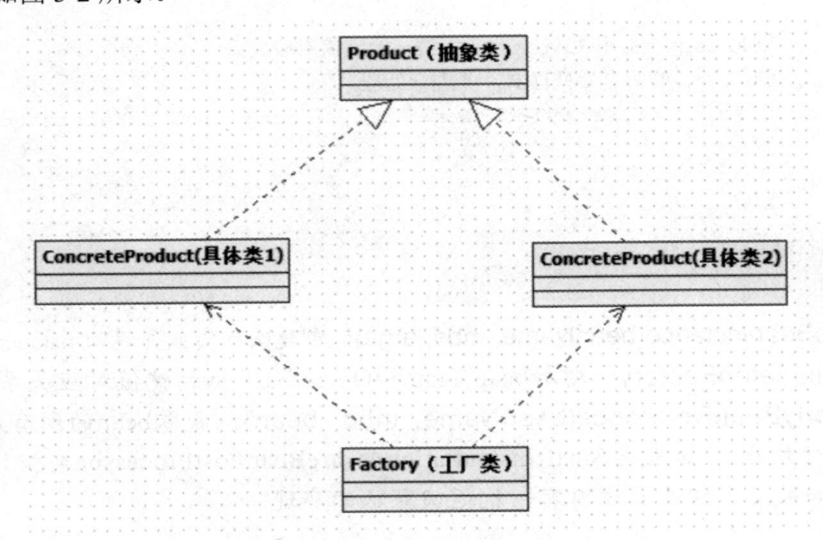

图 3-2 简单工厂模式结构图

在简单工厂模式结构图中包含如下几个角色：

- Factory（工厂角色）：工厂角色即工厂类，它是简单工厂模式的核心，负责实现创建所有产品实例的内部逻辑；工厂类可以被外界直接调用，创建所需的产品对象；在工厂类中提供了静态的工厂方法 factoryMethod()，它的返回类型为抽象产品类型 Product。
- Product（抽象产品角色）：它是工厂类所创建的所有对象的父类，封装了各种产品对象的公有方法，它的引入将提高系统的灵活性，使得在工厂类中只需定义一个通用的工厂方法，因为所有创建的具体产品对象都是其子类对象。

- ConcreteProduct（**具体产品角色**）：它是简单工厂模式的创建目标，所有被创建的对象都充当这个角色的某个具体类的实例。每一个具体产品角色都继承了抽象产品角色，需要实现在抽象产品中声明的抽象方法。

来看一个简单工厂模式的具体例子：

```
/**
 * 描述：交通工具（简单工厂模式）
 * @author Ay
 * @create 2018/1/19
 */
public class SimpleFactoryPattern {

    public static void main(String[] args) {
         //根据需要传入相关的交通工具名称，获取交通工具实例
        Vehicle vehicle = Factory.produce("car");
        vehicle.run();
    }
}
/**
 * 工厂类
 */
class Factory{

    //静态方法，生产交通工具
    public static Vehicle produce(String type){
        Vehicle vehicle = null;
        if(type.equals("car")){
            vehicle = new _Car();
            return vehicle;
        }
        if(type.equals("bus")){
            vehicle = new _Bus();
            return vehicle;
        }
        if(type.equals("bicycle")){
            vehicle = new _Bicycle();
            return vehicle;
        }
        return vehicle;
    }
}

/**
```

```
 * 交通工具（抽象类）
 */
interface Vehicle{

   void run();
}

/**
 * 汽车（具体类）
 */
class Car implements Vehicle{

   @Override
   public void run() {
      System.out.println("car run...");
   }
}

/**
 * 公交车（具体类）
 */
class Bus implements Vehicle{
   @Override
   public void run() {
      System.out.println("bus run...");
   }
}

/**
 * 自行车（具体类）
 */
class Bicycle implements Vehicle{
   @Override
   public void run() {
      System.out.println("bicycle run...");
   }
}
```

上述代码中，交通工具都被抽象为 Vehicle 类，而 Car、Bus、Bicycle 类是 Vehicle 类的实现类，并实现 Vehicle 的 run 方法，打印相关的信息。Factory 是工厂类，类中的静态方法 produce 根据传入的不同的交通工具的类型，生产相关的交通工具，返回抽象产品类 Vehicle。上述代码的类结构如图 3-3 所示。

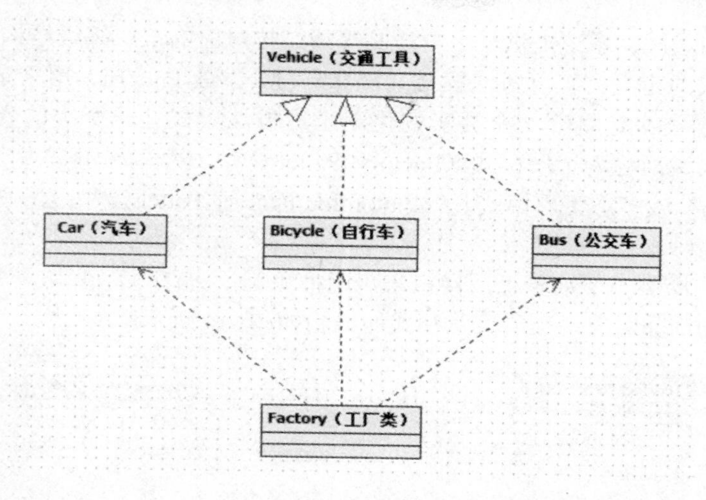

图 3-3　汽车与工厂类结构图

简单工厂模式的主要优点如下：

（1）工厂类包含必要的判断逻辑，可以决定在什么时候创建哪一个产品类的实例，客户端可以免除直接创建产品对象的职责，而仅仅“消费”产品，简单工厂模式实现了对象创建和使用的分离。

（2）客户端无须知道所创建的具体产品类的类名，只需要知道具体产品类所对应的参数即可，对于一些复杂的类名，通过简单工厂模式可以在一定程度减少使用者的记忆量。

简单工厂模式的主要缺点如下：

（1）由于工厂类集中了所有产品的创建逻辑，职责过重，一旦不能正常工作，整个系统都会受到影响。

（2）使用简单工厂模式势必会增加系统中类的个数（引入了新的工厂类），增加了系统的复杂度和理解难度。

（3）系统扩展困难，一旦添加新产品就不得不修改工厂逻辑，在产品类型较多时，有可能造成工厂逻辑过于复杂，不利于系统的扩展和维护。

（4）简单工厂模式由于使用了静态工厂方法，造成工厂角色无法形成基于继承的等级结构。

3.1.5　工厂方法模式详解

在简单工厂模式中只提供一个工厂类，它需要知道每一个产品对象的创建细节，并决定何时实例化哪一个产品类。简单工厂模式最大的缺点是当有新产品要加入到系统中时，必须修改工厂类，需要在其中加入必要的业务逻辑，这违背了“开闭原则”。此外，在简单工厂模式中，所有的产品都由同一个工厂创建，工厂类职责较重，业务逻辑较为复杂，具体产品与工厂类之间的耦合度高，严重影响了系统的灵活性和扩展性，而工厂方法模式则可以很好地解决这一问题。

在工厂方法模式中，不再提供一个统一的工厂类来创建所有的产品对象，而是针对不同的产品提供不同的工厂，系统提供一个与产品等级结构对应的工厂等级结构。

工厂方法模式的定义：工厂方法模式（Factory Method Pattern）用来定义一个用于创建对象的接口，让子类决定将哪一个类实例化。工厂方法模式让一个类的实例化延迟到其子类。工厂方法模式又简称为工厂模式（Factory Pattern），还可称作虚拟构造器模式（Virtual Constructor Pattern）或多态工厂模式（Polymorphic Factory Pattern）。工厂方法模式是一种类创建型模式。

工厂方法模式提供一个抽象工厂接口来声明抽象工厂方法，而由其子类来具体实现工厂方法，创建具体的产品对象。工厂方法模式的结构如图 3-4 所示。

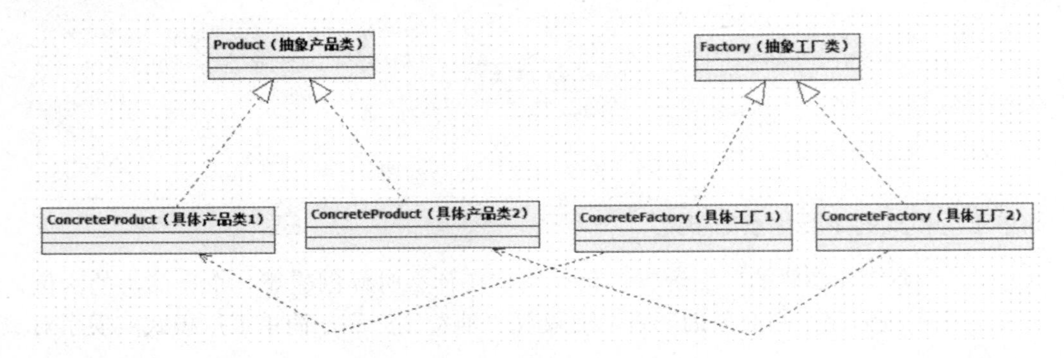

图 3-4　工厂方法模式类结构图

在工厂方法模式结构图中包含如下几个角色。

- Product（抽象产品类）：它是定义产品的接口，是工厂方法模式所创建对象的超类型，也就是产品对象的公共父类。
- ConcreteProduct（具体产品类）：它实现了抽象产品接口，某种类型的具体产品由专门的具体工厂创建，具体工厂和具体产品之间一一对应。
- Factory（抽象工厂类）：在抽象工厂类中，声明了工厂方法（Factory Method），用于返回一个产品。抽象工厂是工厂方法模式的核心，所有创建对象的工厂类都必须实现该接口。
- ConcreteFactory（具体工厂类）：它是抽象工厂类的子类，实现了抽象工厂中定义的工厂方法，并可由客户端调用，返回一个具体产品类的实例。

下面来看一个汽车与工厂实例，具体代码如下：

```
/**
 * 描述：汽车与工厂（工厂方法模式）
 * @author Ay
 * @create 2018/1/19
 */
public class FactoryMethodPattern {
    public static void main(String[] args) throws Exception {
        //生产汽车
        Factory carFactory = new CarFactory();
        Vehicle car = carFactory.produce();
```

```
        car.run();
        //生产公交车
        Factory busFactory = new BusFactory();
        Vehicle bus = busFactory.produce();
        bus.run();
        //生产自行车
        BicycleFactory bicycleFactory = new BicycleFactory();
        Vehicle bicycle = bicycleFactory.produce();
        bicycle.run();
    }
}
/**
 * 抽象工厂类
 */
interface Factory {
    //生产
    Vehicle produce();
}
/**
 * 汽车工厂
 */
class CarFactory implements Factory {
    @Override
    public Vehicle produce() {
        return new Car();
    }
}

/**
 * 公交车工厂
 */
class BusFactory implements Factory {
    @Override
    public Vehicle produce() {
        return new Bus();
    }
}
/**
 * 自行车工厂
 */
class BicycleFactory implements Factory{
    @Override
    public Vehicle produce() {
        return new Bicycle();
    }
```

```
}
/**
*交通工具
*/
interface Vehicle {
    void run();
}

/**
 * 汽车
 */
class Car implements Vehicle {
    @Override
    public void run() {
        System.out.println("car run...");
    }
}

/**
 * 公交车
 */
class Bus implements Vehicle {
    @Override
    public void run() {
        System.out.println("bus run...");
    }
}

/**
 * 自行车
 */
class Bicycle implements Vehicle {
    @Override
    public void run() {
        System.out.println("bicycle run...");
    }
}
```

上述代码中，Vehicle 类是抽象产品类，而 Car、Bus、Bicycle 类是具体产品类，并且实现 Vehicle 类的 run 方法。每一种具体产品类都有一一对应的工厂类 CarFactory、BusFactory、BicycleFactory 等，所有的工厂都有共同的抽象父类 Factory。汽车与工厂具体的类结构如图 3-5 所示。

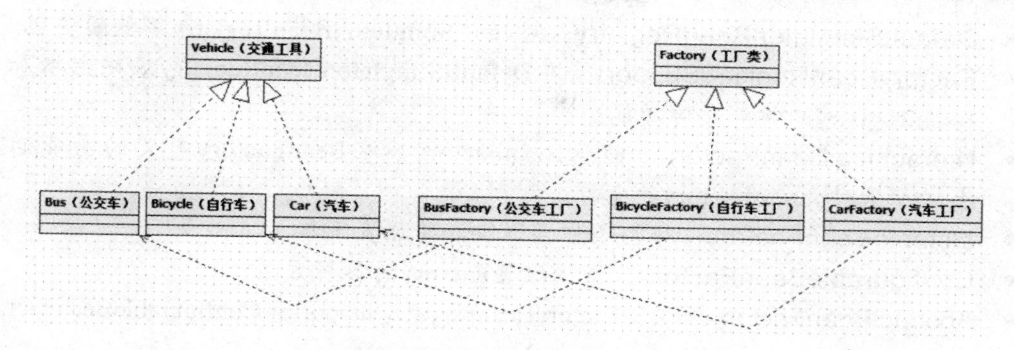

图 3-5 汽车与工厂类结构图

与简单工厂模式相比，工厂方法模式最重要的区别是引入了抽象工厂角色，抽象工厂可以是接口，也可以是抽象类或者具体类。在抽象工厂中声明了工厂方法但并未实现工厂方法，具体产品对象的创建由其子类负责，客户端针对抽象工厂编程，可在运行时再指定具体工厂类，具体工厂类实现了工厂方法，不同的具体工厂可以创建不同的具体产品。

3.1.6 Spring Bean 工厂类详解

Spring 的 bean 工厂类都是存放在 spring-beans-5.0.3.RELEASE.jar 包中的，可以使用 IntelliJ IDEA 查看 Spring 的类结构图，具体方法是：【打开 DefaultListableBeanFactory 类】→【右击】→【Diagrams】→【Show Diagram】→【Java Class Diagrams】，便可打开如图 3-6 所示的 Spring Bean 工厂类结构图。

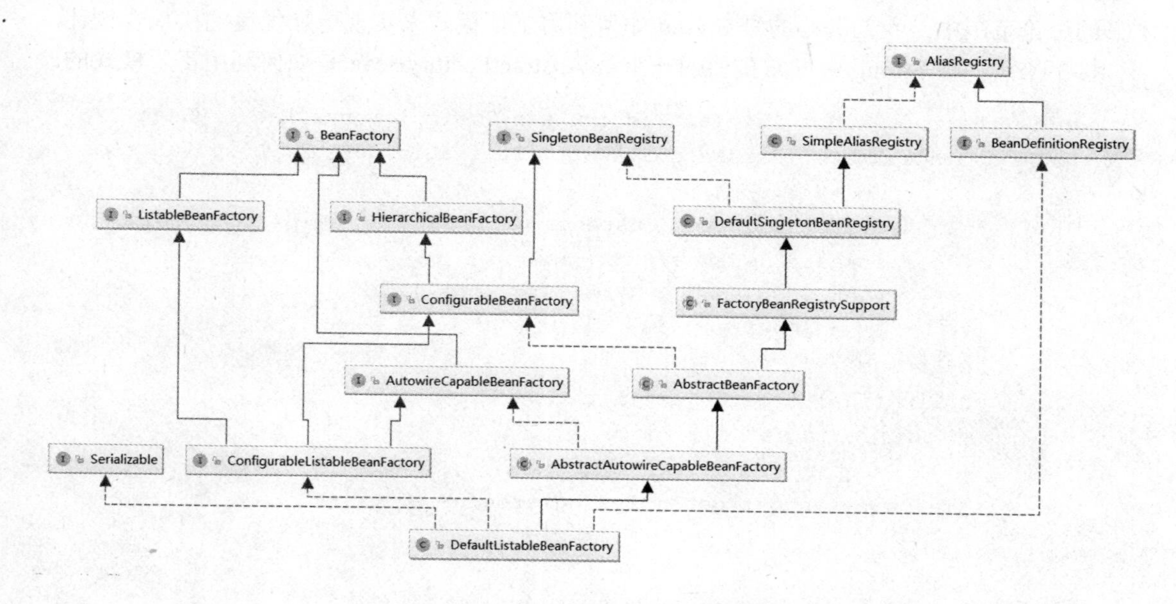

图 3-6 Spring Bean 工厂类结构图

- BeanFactory：定义获取bean及bean的各种属性。
- SimgletonBeanRegistry：定义对单例的注册及获取。

- DefaultSimgletonBeanRegistry：对接口SimgletonBeanRegistry函数的实现。
- FactoryBeanRegistrySupport：在DefaultSimgletonBeanRegistry基础上增加了对BeanRegistry的特殊处理功能。
- HierachicalBeanFactory：继承BeanFactory，在BeanFactory定义的功能的基础上增加了对parentFactory的支持。
- ListableBeanFactory：根据条件获取bean的配置清单。
- ConfigurableBeanFactory：提供配置Factory的各种方法。
- AbstractBeanFactory：综合FactoryBeanRegistrySupport和ConfigurableBeanFactory的功能。
- AutowireCapableBeanFactory：提供创建Bean、自动注入、初始化以及应用bean的后处理器。
- ConfigurableListableBeanFactory：BeanFactory配置清单，指定忽略类型及接口等。
- AbstractAutowireCapableBeanFactory：综合AbstractBeanFactory并对接口Autowire- CapableBeanFactory进行实现。
- DefaultListableBeanFactory：整个Bean加载的核心部分，是Spring注册和加载Bean的默认实现。

Spring 源码中有非常多的地方用到了工厂模式，几乎是无处不见，但是笔者决定以大家最为常用的 Bean 来讲解，使用 Spring 很大程度上是依赖它的对象管理，也就是 IOC 容器对于 Bean 的管理，Spring 的 IOC 容器如何创建和管理 Bean 其实是比较复杂的，它并不在我们本书的讨论范围中，我们关心的是 Spring 如何利用工厂模式来实现更加优良的松耦合设计。

接下来看一下 Spring 中非常重要的一个类 AbstractFactoryBean 是如何利用工厂模式的。

```
public abstract class AbstractFactoryBean<T> implements FactoryBean<T>,
BeanClassLoaderAware, BeanFactoryAware, InitializingBean, DisposableBean {
/**
        * Expose the singleton instance or create a new prototype instance.
        * @see #createInstance()
        * @see #getEarlySingletonInterfaces()
        */
        @Override
        public final T getObject() throws Exception {
            if (isSingleton()) {
                return (this.initialized ?
 this.singletonInstance : getEarlySingletonInstance());
            }
            else {
                return createInstance();
            }
        }
```

```
    protected abstract T createInstance() throws Exception;
}
```

- AbstractFactoryBean：实现FactoryBean类，主要是实现getObject方法，返回bean实例。
- getObject()：如果是单例且已经创建，返回单例模式，未创建调用getEarlySingletonInstance方法，不是单例模式，调用createInstance方法。
- createInstance()：由子类负责创建具体对象。

3.2 Spring AOP

3.2.1 Spring AOP 概述

AOP 为 Aspect Oriented Programming 的缩写，意为面向切面编程，通过预编译方式和运行期动态代理实现程序功能的统一维护的一种技术。AOP 是 OOP 的延续，是软件开发中的一个热点，也是 Spring 框架中的一个重要内容，是函数式编程的一种衍生范型。利用 AOP 可以对业务逻辑的各个部分进行隔离，从而使得业务逻辑各部分之间的耦合度降低，提高程序的可重用性和开发效率。

AOP 主要的功能有日志记录、性能统计、安全控制、事务处理和异常处理等。

3.2.2 Spring AOP 核心概念

首先，来简单理解一下什么是切面，具体请看图 3-7。

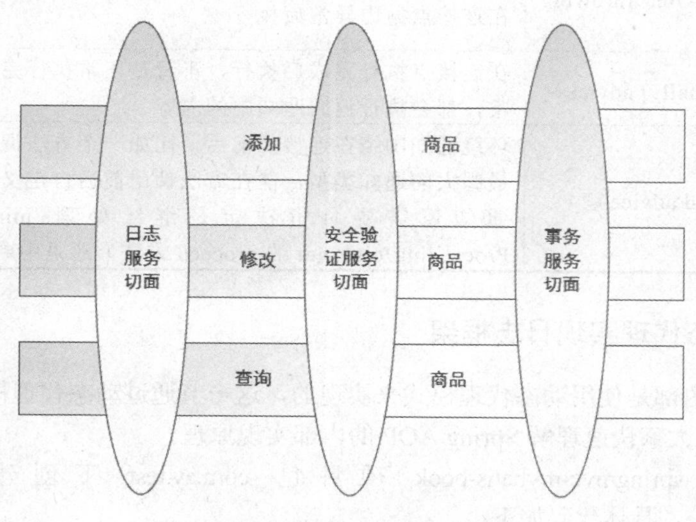

图 3-7 Spring AOP 切面

Spring AOP 切面简单理解就像一把刀，在代码执行过程中，可以随意地插入或拔出。在插入的位置或拔出的位置可以“任意妄为”地做自己喜欢的事情，比如记录日志、控制事务、安全验证服务，等等。

Spring AOP 核心概念如表 3-1 所示。

表 3-1 Spring AOP 核心概念

名　　称	说　　明
横切关注点	对哪些方法进行拦截，拦截后怎么处理，这些关注点称为横切关注点
切面（aspect）	类是对物体特征的抽象，切面就是对横切关注点的抽象
连接点（joinpoint）	指被拦截到的点，因为 Spring 只支持方法类型的连接点，所以在 Spring 中连接点指的就是被拦截到的方法，实际上连接点还可以是字段或构造器
切入点（pointcut）	对连接点进行拦截的定义
通知（advice）	所谓通知指的就是拦截到连接点之后要执行的代码，通知分为前置、后置、异常、最终、环绕通知 5 类
目标对象	代理的目标对象
织入（weave）	将切面应用到目标对象并导致代理对象创建的过程
引入（introduction）	在不修改代码的前提下，引入可以在运行期为类动态地添加一些方法或字段

Spring AOP 通知（advice）分成 5 类，具体如表 3-2 所示。

表 3-2 Advice 通知类型

名　　称	说　　明
前置通知（Before advice）	在连接点前面执行，前置通知不会影响连接点的执行，除非此处抛出异常
正返回通知（After returning advice）	在连接点正常执行完成后执行，如果连接点抛出异常，则不会执行
异常返回通知（After throwing advice）	在连接点抛出异常后执行
后通知（After [finally] advice）	在连接点执行完成后执行，不管是正常执行完成，还是抛出异常，都会执行返回通知中的内容
环绕通知（Around advice）	环绕通知围绕在连接点前后，比如一个方法调用的前后。这是最强大的通知类型，能在方法调用前后自定义一些操作。环绕通知还需要负责决定是继续处理 join point（调用 ProceedingJoinPoint 的 proceed 方法）还是中断执行

3.2.3 JDK 动态代理实现日志框架

Spring AOP 内部是使用动态代理模式来实现的，这一节通过动态代理模式来实现最简单的日志框架，帮助大家快速理解 Spring AOP 的内部实现原理。

首先，在 springmvc-mybatis-book 项目包 com.ay.test 下创建业务接口类 BusinessClassService，具体代码如下：

```
package com.ay.test;
/**
 * 描述：业务类接口
 * @author Ay
```

```
 * @create 2018/04/22
 **/
public interface BusinessClassService {

    void doSomeThing();
}
```

BusinessClassService 业务接口类可以理解为日常开发业务创建的接口类，接口中有一个简单的方法 doSomeThing。然后，开发业务类的实现类 BusinessClassServiceImpl，具体代码如下：

```
package com.ay.test;

/**
 * 描述：业务实现类
 * @author Ay
 * @create 2018/04/22
 **/
public class BusinessClassServiceImpl implements BusinessClassService{

    /**
     * 处理业务
     */
    public void doSomeThing(){
        System.out.println("do something ......");
    }
}
```

实现类 BusinessClassServiceImpl 实现了 BusinessClassService 接口，并实现了 doSomeThing 方法，在方法中打印“do something ......”。

接着，开发日志接口类 MyLogger，具体代码如下：

```
package com.ay.test;
import java.lang.reflect.Method;
/**
 * 描述：日志类接口
 * @author Ay
 * @create 2018/04/22
 **/
public interface MyLogger {
    /**
     * 记录进入方法时间
     */
    void saveIntoMethodTime(Method method);
    /**
     * 记录退出方法时间
     */
    void saveOutMethodTime(Method method);
```

```
}
```

- saveIntoMethodTime：记录进入方法的时间。
- saveOutMethodTime：记录退出方法的时间。

接口类 MyLogger 开发完成之后，用 MyLoggerImpl 类实现它，具体代码如下：

```
package com.ay.test;
import java.lang.reflect.Method;

/**
 * 描述：日志实现类
 * @author Ay
 * @create 2018/04/22
 **/
public class MyLoggerImpl implements MyLogger {

    public void saveIntoMethodTime(Method method) {
        System.out.println("进入" + method.getName()  +
"方法时间为: " + System.currentTimeMillis());
    }

    public void saveOutMethodTime(Method method) {
        System.out.println("退出" + method.getName() + "方法时间为: " +
System.currentTimeMillis());
    }
}
```

MyLoggerImpl 类实现接口 MyLogger，并实现 saveIntoMethodTime 和 saveOutMethodTime 方法，在方法内部打印进入/退出方法的时间。

最后，实现最重要的类 MyLoggerHandler，具体代码如下：

```
package com.ay.test;
import javax.annotation.Resource;
import java.lang.reflect.InvocationHandler;
import java.lang.reflect.Method;

/**
 * 描述：日志类 Handler
 * @author Ay
 * @create 2018/04/22
 **/
public class MyLoggerHandler implements InvocationHandler {

    //原始对象
    private Object objOriginal;
    //这里很关键
```

```
    private MyLogger myLogger = new MyLoggerImpl();

    public MyLoggerHandler(Object obj){
        super();
        this.objOriginal = obj;
    }

    public Object invoke(Object proxy, Method method, Object[] args) throws
Throwable {
        Object result = null;
        //日志类的方法：保存进入方法的时间
        myLogger.saveIntoMethodTime(method);
        //调用代理类方法
        result = method.invoke(this.objOriginal ,args);
        //日志类方法：保存退出方法的时间
        myLogger.saveOutMethodTime(method);
        return result;
    }
}
```

- InvocationHandler：该接口中仅定义了一个方法：public Object invoke(Object obj, Method method, Object[] args)，在使用时，第一个参数obj一般是指代理类，method是被代理的方法，args为该方法的参数数组。这个抽象方法在代理类中动态实现。

所有的代码开发完成之后，开发测试类 MyLoggerTest 进行测试，具体代码如下：

```
package com.ay.test;
import java.lang.reflect.Proxy;
/**
 * 描述：测试类
 * @author Ay
 * @create 2018/04/22
 **/
public class MyLoggerTest {

    public static void main(String[] args) {
        //实例化真实项目中业务类
        BusinessClassService businessClassService = new
BusinessClassServiceImpl();
        //日志类的 handler
        MyLoggerHandler myLoggerHandler = new
MyLoggerHandler(businessClassService);
        //获得代理类对象
        BusinessClassService businessClass = (BusinessClassService)
        Proxy.newProxyInstance(businessClassService.getClass().
```

```
        getClassLoader(),businessClassService.getClass().getInterfaces(),
        myLoggerHandler);
        //执行代理类方法
        businessClass.doSomeThing();
    }
}
```

- Proxy.newProxyInstance：该类即为动态代理类，static Object newProxyInstance (ClassLoader loader, Class[] interfaces, InvocationHandler h)，返回代理类的一个实例，返回后的代理类可以当作被代理类使用。在Proxy.newProxyInstance方法中，共有以下三个参数：
  - ClassLoader loader：targetObject.getClass().getClassLoader()目标对象通过getClass方法获取类的所有信息后，调用getClassLoader()方法来获取类加载器。获取类加载器后，可以通过这个类型的加载器，在程序运行时，将生成的代理类加载到JVM即Java虚拟机中，以便运行时需要。
  - Class[] interfaces：targetObject.getClass().getInterfaces()获取被代理类的所有接口信息，以便于生成的代理类可以具有代理类接口中的所有方法。
  - InvocationHandler h：使用动态代理是为了更好地扩展，比如在方法之前做什么操作，之后做什么操作，这个时候这些公共的操作可以统一交给代理类去做。此时需要调用实现了InvocationHandler 类的一个回调方法。

运行测试类的 main 方法，便可以在 IntelliJ IDEA 控制台查看打印信息，具体信息如下：

```
进入 doSomeThing 方法时间为：1524385006965
do something ......
退出 doSomeThing 方法时间为：1524385006966
```

以上就是利用动态代理模式实现简单的日志框架，具体的结构如图 3-8 所示。

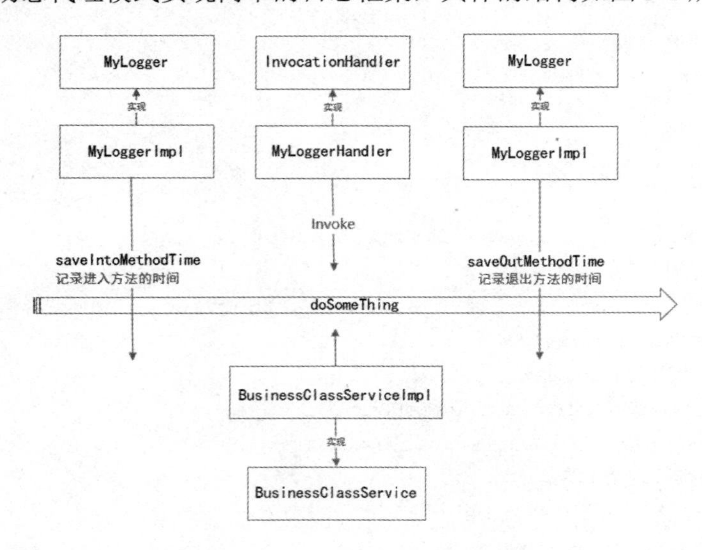

图 3-8 Spring AOP 实现日志框架的结构图

这里总结一下，JDK 动态代理的一般实现步骤如下：

（1）创建一个实现 InvocationHandler 接口的类 MyLoggerHandler，它必须实现 invoke 方法。

（2）创建被代理的类 BusinessClassService 以及接口 BusinessClassServiceImpl。

（3）调用 Proxy 的静态方法 newProxyInstance，创建一个代理类。

（4）通过代理类调用方法。

3.2.4 Spring AOP 实现日志框架

使用 Spring AOP 的注解方式实现日志框架是非常简单的。首先，在配置文件 spring-mvc.xml 中添加配置，具体代码如下：

```
<aop:aspectj-autoproxy proxy-target-class="true">
```

- </aop:aspectj-autoproxy>：声明自动为Spring容器中那些配置@aspectJ切面的bean创建代理，织入切面。<aop:aspectj-autoproxy />有一个proxy-target-class属性，默认为false，表示使用JDK动态代理织入增强，当配置poxy-target-class为true时，表示使用CGLib动态代理技术织入增强。不过即使设置proxy-target-class为false，如果目标类没有声明接口，则Spring将自动使用CGLib动态代理。

配置添加完成之后，要定义一个切面 LogInterceptor，具体代码如下：

```
package com.ay.proxy;
import org.aspectj.lang.annotation.After;
import org.aspectj.lang.annotation.Aspect;
import org.aspectj.lang.annotation.Before;
import org.springframework.stereotype.Component;

/**
 * 描述：日志拦截类（切面）
 * @author Ay
 * @create 2018/04/22
 **/
@Aspect
@Component
public class LogInterceptor {

    @Before(value = "execution(* com.ay.controller.*.*(..))")
    public void before(){
        System.out.println("进入方法时间为:" + System.currentTimeMillis());
    }

    @After(value = "execution(* com.ay.controller.*.*(..))")
    public void after(){
        System.out.println("退出方法时间为:" + System.currentTimeMillis());
```

```
    }
}
```

- **@Aspect**：标识LogInterceptor类为一个切面，供容器读取。
- **@Before**：在所拦截方法执行之前执行before方法。
- **@After**：在所拦截方法执行之后执行after方法。
- **@Around**：可以同时在所拦截方法的前后执行一段逻辑。
- **execution切入点指示符**：com.ay.controller.*.*(..)表示在controller包中定义的任意方法的执行。execution切入点指示符执行表达式的格式如下：

```
execution (modifiers-pattern? ret-type-pattern declaring-type-pattern?
name-pattern (param-pattern) throws-pattern?)
```

翻译为：

```
execution(方法修饰符  方法返回值  方法所属类 匹配方法名 (方法中的形参表) 方法申明抛出的
异常)
```

其中黑色字体部分不能省略，各部分都支持通配符“*”来匹配全部。特殊的为形参表部分，其支持以下两种通配符：

- "*"：代表一个任意类型的参数。
- ".."：代表零个或多个任意类型的参。

例如：

- ()：匹配一个无参方法。
- (..)：匹配一个可接受任意数量参数和类型的方法。
- (*)：匹配一个接受一个任意类型参数的方法。
- (*，Integer)：匹配一个接受两个参数的方法，第一个可以为任意类型，第二个必须为Integer。

下面举一些execution的使用实例，具体内容见表3-3。

表3-3 切入点表达式实例

切入点表达式	说 明
execution(public * * (..))	匹配所有目标类的 public 方法，第一个*为返回类型，第二个*为方法名
execution(* save* (..))	匹配所有目标类以 save 开头的方法，第一个*代表返回类型
execution(**product(*,String))	匹配目标类所有以 product 结尾的方法，并且其方法的参数表第一个参数可为任意类型，第二个参数必须为 String
execution(* aop_part.Demo1.service.*(..))	匹配 service 接口及其实现子类中的所有方法
execution(* aop_part.*(..))	匹配 aop_part 包下的所有类的所有方法，但不包括子包
execution(* aop_part..*(..))	匹配 aop_part 包下的所有类的所有方法，包括子包。（当".."出现在类名中时，后面必须跟“*”,表示包、子孙包下的所有类）

（续表）

切入点表达式	说　明
execution(*aop_part..*.*service.find*(..))	匹配 aop_part 包及其子包下的所有后缀名为 service 的类中所有方法名必须以 find 为前缀的方法
execution(*foo(String,int))	匹配所有方法名为 foo，且有两个参数，其中，第一个参数的类型为 String，第二个参数的类型为 int
execution(* foo(String,..))	匹配所有方法名为 foo，且至少含有一个参数，并且第一个参数为 String 的方法（后面有任意个类型不限的形参）

切面类 LogInterceptor 开发完成之后，重新启动 springmvc-mybatis-book 项目，项目成功启动后，在浏览器输入网址：http://localhost:8080/user/findAll，便可以在 IntelliJ IDEA 开发工具的控制台看到如下的打印信息：

```
进入方法时间为:1524411433320
id: 1
name: 阿毅
id: 2
name: 阿兰
退出方法时间为:1524411434036
```

3.2.5　静态代理与动态代理模式

Spring AOP 使用的是动态代理模式，所以有必要简单学习一下动态代理模式。

代理模式定义如下：

（1）代理模式给某一个对象提供一个代理或占位符，并由代理对象来控制对原对象的访问。

（2）代理模式是一种对象结构型模式。在代理模式中引入了一个新的代理对象，代理对象在客户端对象和目标对象之间起到中介的作用，它去掉客户不能看到的内容和服务或者增添客户需要的额外的新服务。

（3）代理模式的结构比较简单，其核心是代理类，为了让客户端能够一致性地对待真实对象和代理对象，在代理模式中引入了抽象层，代理模式结构如图 3-9 所示。

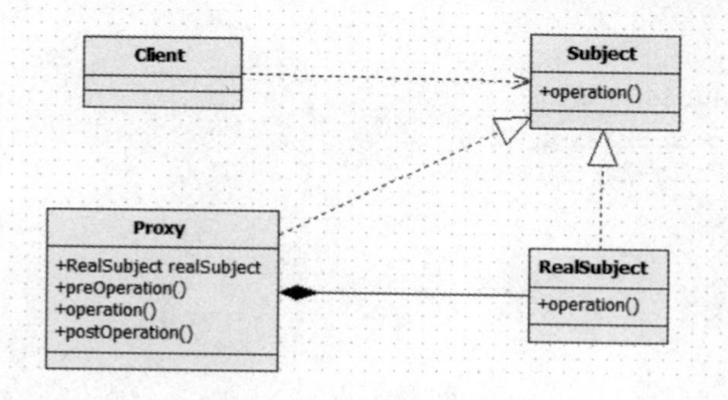

图 3-9　代理模式类结构图

由图 3-9 可知，代理模式包含如下三个角色：

- Subject（**抽象主题角色**）：它声明了真实主题和代理主题的共同接口，这样一来在任何使用真实主题的地方都可以使用代理主题，客户端通常需要针对抽象主题角色进行编程。
- Proxy（**代理主题角色**）：它包含了对真实主题的引用，从而可以在任何时候操作真实主题对象；在代理主题角色中提供一个与真实主题角色相同的接口，以便在任何时候都可以替代真实主题；代理主题角色还可以控制对真实主题的使用，负责在需要的时候创建和删除真实主题对象，并对真实主题对象的使用加以约束。通常，在代理主题角色中，客户端在调用所引用的真实主题操作之前或之后还需要执行其他操作，而不仅仅是单纯调用真实主题对象中的操作。
- RealSubject（**真实主题角色**）：它定义了代理角色所代表的真实对象，在真实主题角色中实现了真实的业务操作，客户端可以通过代理主题角色间接调用真实主题角色中定义的操作。

静态代理模式具体的代码如下：

```
package com.ay.test;
/**
 * 描述：客户端类
 * @author Ay
 * @create 2018/04/22
 **/
public class ProxyPattern{
    public static void main(String[] args) {
        //为每个 RealSubject 创建代理类 Proxy
        Proxy proxy = new Proxy(new RealSubject());
        proxy.operation();
    }
}
/**
 * 描述：抽象主题类
 * @author Ay
 * @create 2018/04/22
 **/
abstract class Subject {
    abstract void operation();
}

/**
 * 描述：具体主题类
 * @author Ay
 * @create 2018/04/22
 **/
```

```
class RealSubject extends Subject{

    void operation() {
        System.out.println("operation ......");
    }
}
/**
 * 描述：代理类
 * @author Ay
 * @create 2018/04/22
 **/
class Proxy extends Subject{

    private Subject subject;

    public Proxy(Subject subject){
        this.subject = subject;
    }

    void operation() {
        //前置处理
        this.preOperation();
        //具体操作
        subject.operation();
        //后置处理
        this.postOperation();
    }

    void preOperation(){
        System.out.println("pre operation......");
    }

    void postOperation(){
        System.out.println("post operation......");
    }
}
```

上面介绍的代理模式也被称为“静态代理模式”，这是因为在编译阶段就要为每个 RealSubject 类创建一个 Proxy 类，当需要代理的类很多时，就会出现大量的 Proxy 类，所以可以使用 JDK 动态代理解决这个问题。关于 JDK 动态代理实例，大家可以参考 3.2.3 节，JDK 动态代理的实现原理是动态创建代理类并通过指定类加载器加载，然后在创建代理对象时将 InvokerHandler 对象作为构造参数传入。当调用代理对象时，会调用 InvokerHandler.invoke()方法，并最终调用真正业务对象的相应方法。

3.3 思考与练习

1. 简述 Spring IOC 和 DI。
2. 请列举出常用的声明 Bean 注解。
3. 注入 Bean 注解有哪些?
4. 简述@Resource 和@Autowired 注解的区别。
5. 简单描述单例模式的定义。
6. 实现一个单例模式，需要哪三个步骤？
7. 动手实现一个饿汉模式的单例。
8. 动手实现一个懒汉模式的单例。
9. 双重校验锁会出现什么问题，具体如何解决？
10. 简单描述一下什么是指令重排。
11. 简单画下简单工厂模式结构图。
12. 简单画下工厂方法模式结构图。
13. 简单工厂模式相比工厂方法模式最重要的区别是什么？
14. 简述 Spring AOP 的含义。
15. Spring AOP 通知类型有哪些？
16. 简述 JDK 动态代理的实现步骤。
17. 静态代理的缺点是什么？
18. 切入点表达式 execution(* save* (..))的含义是什么？

第4章

MyBatis 映射器与动态 SQL

本章主要介绍 MyBatis 常用的映射器元素、动态 SQL 元素、MyBatis 注解配置和关联映射。

4.1 MyBatis 映射器

MyBatis 框架包括两种类型的 XML 文件，一类是配置文件，即 mybatis-config.xml；另外一类是映射文件，例如 XXXMapper.xml 等。在 MyBatis 的配置文件 mybatis-config.xml 包含了<mappers></mappers>节点，这里就是 MyBatis 映射器。

4.1.1 映射器的主要元素

MyBatis 提供了丰富的映射器元素，可以实现许多功能，这些元素的含义如表 4-1 所示。

表 4-1 映射器元素

元素名称	描　　述
select	映射查询语句
insert	映射插入语句
update	映射更新语句
delete	映射删除语句
sql	可以被其他语句引用的可重用语句块
resultMap	用来描述如何从数据库结果集中来加载对象
cache	给定命名空间的缓存配置
cache-ref	其他命名空间缓存配置的引用

接下来详细讨论映射器中主要元素的用法。

4.1.2 select 元素

select 元素是 MyBatis 中最常用的元素之一，select 元素可以从数据库中读取数据，组装数据给业务人员。比如可以在配置文件 AyUserMapper.xml 中使用 select 元素，根据用户 ID 查询 ay_user 表（2.4 节已创建）中的具体用户，具体代码如下：

```
    <select id="findById" parameterType="String"
resultType="com.ay.model.AyUser">
         SELECT * FROM ay_user
         WHERE id = #{id}
    </select>
```

这个语句被称为 findById，接受一个 String 类型的参数，并返回一个 User 类型的对象。参数#{id}是告诉 MyBatis 创建一个预处理语句参数。通过 JDBC，这样的一个参数在 SQL 中会由一个“？”来标识，并被传递到一个新的预处理语句中。上面的 SQL 语句执行时会生成如下的 JDBC 代码：

```
String findById =  "SELECT * FROM ay_user WHERE id = ? "
PreparedStatement ps = conn.prepareStatement(findById);
ps.setString(1,id);
```

接口 AyUserDao 中定义的方法如下：

```
AyUser findById(String id);
```

select 元素提供了很多配置属性，具体如表 4-2 所示。

表 4-2 select 元素的属性

属性名称	描　　述
id	它和 mapper 的命名空间组合起来是唯一的，id 的值和 DAO 接口的方法名一致。如果不唯一或者不一致，MyBatis 将抛出异常
parameterType	将会传入语句参数类的全名码或者别名，这个属性是可选的，因为 MyBatis 可以通过 TypeHandler 推断出具体传入语句的参数，默认值为 unset。可以选择 JavaBean、Map 等复杂的参数类型传递给 SQL
parameterMap	即将废弃的元素，不再讨论
resultType	从语句中返回期望类型的类的完全限定名或别名。注意如果是集合的情形，应该是集合可以包含的类型，而不能是集合本身。返回时可以使用 resultType 或者 resultMap，但不能同时使用。结果集将通过 JavaBean 的规范映射或定义为 int、double、float 等参数
resultMap	它是映射集的引用，将执行强大的映射功能，可以使用 resultType 或者 resultMap 其中的一个，resultMap 可以给予我们自定义映射规则的机会
flushCache	它的作用是调用 SQL 后，是否要求 MyBatis 清空之前查询的本地缓存和二级缓存，取值为 false/true，默认为 false

（续表）

属性名称	描　　述
useCache	启动二级缓存的开关，取值 true/false，默认值为 true
timeout	设置超时参数，等超时的时候抛出异常，单位为秒
fetchSize	获取记录的总条数设定
statementType	告诉 MyBatis 使用哪个 JDBC 的 Statement 工作，取值为 STATEMENT、PREPARED 或者 CALLABLE。默认为 PREPARED
resultSetType	它的值包括 FORWARD_ONLY（游标允许向前访问）|SCROLL_SENSITIVE（双向滚动，并及时跟踪数据库更新，以便更改 resultSet 中的数据）|SCROLL_INSENSITIVE（双向滚动，但不及时跟踪数据库更新，数据库里的数据修改，并不在 resultSet 中反应过来）
databaseId	如果设置了 databaseIdProvider，MyBatis 会加载所有的不带 databaseId 或匹配当前 databaseId 的语句，如果带或者不带的语句都有，则不带的会被忽略
resultOrdered	这个设置仅针对嵌套结果 select 语句：如果设置为 true，是指假设包含了嵌套结果集或者分组了，当返回一个主结果行的时候，就不会发生对前面结果集引用的情况。这就使得获取嵌套的结果集时不至于导致内存不够用。默认为 false
resultSets	适应于多个结果集的情况，它将列出执行 SQL 后每个结果集的名称，每个名称之间用逗号分隔。使用比较少

下面再来看几个 select 元素的例子：

```
    //实例 1：通过名称查询用户
    <select id="findByName" parameterType="String" resultType=
"com.ay.model.AyUser">
        SELECT * FROM ay_user
        WHERE name = #{name}
    </select>
    //实例 2：通过名称查询用户个数
    <select id="countByName" parameterType="String" resultType="int">
        SELECT count(*) FROM ay_user
        WHERE name = #{name}
    </select>
```

对应的 AyUserDao 接口如下：

```
List<AyUser> findByName(String name);
int countByName(String name);
```

4.1.3　insert 元素

insert 元素用来映射 DML 语句，MyBatis 会在执行插入之后返回一个整数，来表示进行操作后插入的记录数，insert 元素的属性和 select 元素属性差不多，特有的属性如表 4-3 所示。

表 4-3 insert 元素的属性

属性名称	描　述
useGeneratedKeys	令 MyBatis 使用 JDBC 的 useGeneratedKeys 方法来获取由数据库内部生成的主键，例如 MySQL 和 SQL Server 自动递增字段，Oracle 的序列等，使用它时必须给 keyProperty 或者 keyColumn 赋值
keyProperty	表示以哪个列作为属性的主键，不能和 keyColumn 同时使用
keyColumn	指明第几列是主键，不能和 keyProperty 同时使用，只接受整形参数

下面来看几个例子：

```
//实例 1：插入用户数据
<insert id="insert" parameterType="com.ay.model.AyUser">
   INSERT INTO ay_user(id, name, password) VALUE (#{id}, #{name},
#{password});
</insert>
//实例 2：插入用户数据，主键自增
<insert id="insert" useGeneratedKeys="true"
keyProperty="id" parameterType="com.ay.model.AyUser">
   INSERT INTO ay_user(name, password) VALUE (#{name}, #{password});
</insert>
```

对应的 AyUserDao 接口如下：

```
int insert(AyUser ayUser);
```

4.1.4 selectKey 元素

可以使用 keyProperty 属性指定哪个是主键字段，同时使用 useGeneratedKeys 属性告诉 MyBatis 这个主键是否使用数据库内置策略生成。实际工作中往往并非想象中的那么简单，比如希望通过原有主键 id + 1 的方式生成主键 id，具体代码如下：

```
<insert id="insert" useGeneratedKeys="true"
keyProperty="id" parameterType="com.ay.model.AyUser">
   <selectKey keyProperty="id" resultType="int" order="BEFORE">
      SELECT MAX(id) + 1 AS id FROM ay_user
   </selectKey>
      INSERT INTO ay_user(id, name, password) VALUE (#{id}, #{name},
#{password});
</insert>
```

selectKey 元素描述如下：

```
<selectKey
  keyProperty="id"
  resultType="int"
  order="BEFORE"
  statementType="PREPARED">
```

selectKey 元素的属性如表 4-4 所示。

表 4-4　selectKey 元素的属性

属性名称	描　述
keyProperty	selectKey 语句结果应该被设置的目标属性，一般会设置到 id 属性，如果希望得到多个生成的列，可以用逗号分隔属性名称列表
keyColumn	匹配属性的返回结果集中的列名称。如果希望得到多个生成的列，可以用逗号分隔属性名称列表
resultType	结果的类型
order	可以设置为 BEFORE 或者 AFTER。设置为 BEFORE，那么它会首先选择主键，设置 keyProperty，然后执行插入语句。如果设置为 AFTER，那么先执行插入语句，然后是 selectKey 元素
statementType	与之前的 select 元素属性相同，具体请看 4.1.2 节的内容

4.1.5　update 元素

update 元素用来映射 DML 语句，主要用来更新数据库中的数据，MyBatis 会在执行更新操作之后返回一个整数，来表示进行操作后更新的记录数，update 元素属性和 select 元素属性差不多，它特有的属性如表 4-5 所示。

表 4-5　update 元素属性

属性名称	描　述
useGeneratedKeys	令 MyBatis 使用 JDBC 的 useGeneratedKeys 方法来获取由数据库内部生成的主键，例如 MySQL 和 SQL Server 自动递增字段，Oracle 的序列等，使用它时必须给 keyProperty 或者 keyColumn 赋值
keyProperty	表示以哪个列作为属性的主键，不能和 keyColumn 同时使用
keyColumn	指明第几列是主键，不能和 keyProperty 同时使用，只接受整形参数

下面来看一个具体的实例：

```
<update id="update" parameterType="com.ay.model.AyUser">
    UPDATE ay_user SET
    name = #{name},
    password = #{password}
    WHERE id = #{id}
</update>
```

对应的 AyUserDao 接口如下：

```
int update(AyUser ayUser);
```

4.1.6　delete 元素

delete 元素用来映射 DML 语句，主要用来删除数据库中的数据，MyBatis 会在执行删除操作之后返回一个整数，来表示进行删除后更新的记录数，delete 元素属性和 select 元素属性差不多，它特有的属性如表 4-6 所示。

表 4-6 delete 元素属性

属性名称	描　述
useGeneratedKeys	令 MyBatis 使用 JDBC 的 useGeneratedKeys 方法来获取由数据库内部生成的主键，例如 MySQL 和 SQL Server 自动递增字段，Oracle 的序列等，使用它时必须给 keyProperty 或者 keyColumn 赋值
keyProperty	表示以哪个列作为属性的主键，不能和 keyColumn 同时使用
keyColumn	指明第几列是主键，不能和 keyProperty 同时使用，只接受整形参数

下面来看几个具体的实例：

```
//实例 1：根据 id 删除记录
<delete id="delete" parameterType="int">
    DELETE FROM ay_user
    WHERE id = #{id}
</delete>
//实例 2：根据 name 删除记录
<delete id="deleteByName" parameterType="String">
        DELETE FROM ay_user
        WHERE name = #{name}
</delete>
```

4.1.7 sql 元素

sql 元素可以被用来定义可重用的 SQL 代码段，可以包含在其他语句中。它可以被静态地（在加载参数时）参数化，比如有一条 SQL 语句需要查询十几个字段映射到 JavaBean 中去，而其他的 SQL 语句也有类似的需求，重复写这些字段显然不合理，那么就可以用 sql 元素对这些字段进行“封装”，以达到重复使用。

下面来看几个具体的实例：

```
<sql id="userField">
    a.id as "id",
    a.name as "name",
    a.password as "password"
</sql>
<!-- 获取所有用户 -->
<select id="findAll" resultType="com.ay.model.AyUser">
    Select
        //使用 refid 进行引用
    <include refid="userField"/>
    from ay_user a
</select>
```

可以很方便地使用 include 元素的 refid 属性进行引用，还可以使用定制参数来使用 sql 元素，具体代码如下：

```
<sql id="userField">
        //注意：这里使用$符号而不是#符号，否则程序出现异常
     ${prefix}.id as "id",
     ${prefix}.name as "name",
     ${prefix}.password as "password"
</sql>
<!-- 获取所有用户 -->
<select id="findAll" resultType="com.ay.model.AyUser">
   select
   <include refid="userField">
      <property name="prefix" value="a"/>
   </include>
   from ay_user a
</select>
```

4.1.8　#与$区别

4.1.7 节中的 SQL 语句，使用${prefix}而不使用#{ prefix }，它们之间的区别是：

（1）# {}将传入的数据都当成一个字符串，会对自动传入的数据加一个双引号，具体示例如下：

```
order by #{id}
//如果 id 传入 11，则 sql 解析成：
order by "11"
```

（2）${}将传入的数据直接显示生成在 sql 中，具体示例如下：

```
order by #{id}
//如果 id 传入 11，则 sql 解析成：
order by 11
```

（3）#方式能够很大程度防止 sql 注入，$方式无法防止 sql 注入。

综上所述，一般建议采用#，而不是$。

4.1.9　resultMap 结果映射集

resultMap 结果映射集是 Mybatis 中重要的元素，它的作用是告诉 MyBatis 从结果集中取出的数据转换为开发者所需要的对象。resultMap 元素还包含其他的元素，具体如下：

```
<resultMap>
<constructor>              /*用来将查询结果作为参数注入到实例的构造方法中*/
      <idArg/>             /*标记结果作为 ID*/
      <arg/>               /*标记结果作为普通参数*/
```

```
</constructor>
<id/>                        /*一个 ID 结果，标记结果作为 ID*/
<result/>                    /*一个普通结果，JavaBean 的普通属性或字段*/
<association>                /*关联其他的对象*/
</association>
<collection>                 /*关联其他的对象集合*/
</collection>
<discriminator>              /*鉴别器，根据结果值进行判断，决定如何映射*/
      <case></case>          /*结果值的一种情况，将对应一种映射规则*/
</discriminator>
</resultMap>
```

先来看一个具体的示例，代码如下：

```
<sql id="userField">
   ${prefix}.id as "id",
   ${prefix}.name as "name",
   ${prefix}.password as "password"
</sql>
<resultMap id="userMap" type="com.ay.model.AyUser">
   <id property="id" column="id"/>
   <result property="name" column="name"/>
   <result property="password" column="password"/>
</resultMap>

<select id="findAll" resultMap="userMap">
   select
   <include refid="userField">
      <property name="prefix" value="a"/>
   </include>
   from ay_user a
</select>
```

使用 POJO 存储结果集是最常用的方式，也是推荐的方式。resultMap 元素的属性 id 代表这个 resultMap 的标识，type 代表需要映射的 POJO。<id/>元素表示对象的主键，property 代表 POJO 的属性名称，column 代表数据库 SQL 的列名，这样 POJO 和数据库 SQL 的结果就一一对应起来。

result 和 id 两个元素共有的属性如表 4-7 所示。

表 4-7 result 和 id 元素的属性

属性名称	描　述
property	令 MyBatis 使用 JDBC 的 useGeneratedKeys 方法来获取由数据库内部生成的主键，例如 MySQL 和 SQL Server 自动递增字段，Oracle 的序列等，使用它时必须给 keyProperty 或者 keyColumn 赋值
column	对应 SQL 的列

（续表）

属性名称	描　述
javaType	配置 Java 的类型，可以是特定的类完全限定名或者 MyBatis 上下文的别名
jdbcType	配置数据库类型
typeHandler	类型处理器，允许我们用特定的处理器来覆盖 MyBatis 默认的处理器。这要制定 jdbcType 和 javaType 相互转化的规则

其他的元素，比如 collection、association、discriminator 等元素，将会在后续的章节进行描述。

4.2　动态 SQL

4.2.1　动态 SQL 概述

在项目开发过程中，经常需要根据不同的条件拼接 SQL 语句，而 MyBatis 提供了对 SQL 语句动态的组装能力。MyBatis 采用功能强大的基于 OGNL 的表达式来完成动态 SQL。

常用的动态 SQL 元素如表 4-8 所示。

表 4-8　动态 SQL 元素

属性名称	描　述
if	单条件分支判断语句
choose、when 和 otherwise	多条件分支判断语句，相当于 Java 中的 case when 语句
trim, where, set	用于处理 SQL 拼装问题，辅助元素
foreach	循环语句

4.2.2　if 元素

if 元素主要用来做判断语句，比如要按照名称 name 查询相关的用户，但是 name 参数是可填可不填的条件，如果用户没有填写 name 参数，就不要使用它作为查询条件。具体看下面的示例：

```
<sql id="userField">
    a.id as "id",
    a.name as "name",
    a.password as "password"
</sql>
<resultMap id="userMap" type="com.ay.model.AyUser">
    <id property="id" column="id"/>
    <result property="name" column="name"/>
    <result property="password" column="password"/>
</resultMap>
```

```
    //通过用户名 name 和密码 password 查询用户
    <select id="findByNameAndPassword" parameterType="String"
resultMap="userMap">
        SELECT
        <include refid="userField"></include>
        from ay_user a
        WHERE 1 = 1
        <if test="name != null and name != ''">
            and name = #{name}
        </if>
        <if test="password != null and password != ''">
            and password = #{password}
        </if>
    </select>
```

对应的 AyUserDAO 接口代码如下：

```
    List<AyUser> findByNameAndPassword(@Param("name") String name,
    @Param("password")String password);
```

if 标签常常与 test 属性联合使用且是必选属性。上述代码中，判断 name 或者 password 参数是否为空，如果不为空，拼凑 SQL 语句进行查询。如果为空，则忽略。

4.2.3 choose、when、otherwise 元素

与 if 元素的二重选择相比，choose、when、otherwise 元素提供三重选择，有点类似 switch..case..default 语句，具体示例代码如下：

```
    <sql id="userField">
        a.id as "id",
        a.name as "name",
        a.password as "password"
    </sql>
    <resultMap id="userMap" type="com.ay.model.AyUser">
        <id property="id" column="id"/>
        <result property="name" column="name"/>
        <result property="password" column="password"/>
    </resultMap>
    //通过名称 name 和密码 password 查询用户
    <select id="findByNameAndPassword" parameterType="String"
resultMap="userMap">
        SELECT
        <include refid="userField"></include>
        from ay_user a
        WHERE 1 = 1
        <choose>
            <when test="name != null and name != ''">
```

```
            and name = #{name}
         </when>
         <when test="password != null and password != ''">
            and password = #{password}
         </when>
         <otherwise>
               ORDER BY id DESC
         </otherwise>
      </choose>
   </select>
```

<choose>标签里可以包含多个<when>标签，<otherwise>标签是可选的并不是必填选项，比如下面的代码：

```
   <select id="findByNameAndPassword" parameterType="String"
resultMap="userMap">
      SELECT
      <include refid="userField"></include>
      from ay_user a
      WHERE 1 = 1
      <choose>
         <when test="name != null and name != ''">
            and name = #{name}
         </when>
         <when test="password != null and password != ''">
            and password = #{password}
         </when>
      </choose>
   </select>
```

4.2.4　trim、where、set 元素

trim 是更灵活地用来去处多余关键字的标签，它可以实现 where 和 set 的效果。具体运用看下面的示例：

```
   <select id="findByNameAndPassword" parameterType="String"
resultMap="userMap">
      SELECT
      <include refid="userField"></include>
      from ay_user a
      <trim prefix="WHERE"  prefixOverrides="AND">
        <if test="name != null and name != ''">
            and name = #{name}
        </if>
        <if test="password != null and password != ''">
           and password = #{password}
```

```
        </if>
    </trim>
</select>
```

假如 name 和 password 字段都不为空，上面的代码相当于如下的 SQL 语句：

```
SELECT
a.id as "id",
a.name as "name",
a.password as "password"
from ay_user a
WHERE name = #{name} and password = #{password}
```

再来看另外一个示例：

```
<update id="update" parameterType="com.ay.model.AyUser">
    UPDATE ay_user
    <trim prefix="SET" suffixOverrides=",">
        <if test="name != null and name != ''">
            name = #{name},
        </if>
        <if test="password != null and password != ''">
            password = #{password},
        </if>
    </trim>
    WHERE id = #{id}
</update>
```

假如 name 和 password 字段都不为空，上面的代码相当于如下的 SQL 语句：

```
UPDATE ay_user
SET
name = #{name},
password = #{password}
WHERE id= #{id}
```

trim 元素的属性如表 4-9 所示。

表 4-9　trim 的属性元素

属性名称	描　　述
prefix	表示在 trim 标签包裹的部分前面添加内容。注意：是在没有 prefixOverrides 和 suffixOverrides 属性的情况下
prefixOverrides	有 prefix 属性的情况下，prefixOverrides 属性表示去掉 SQL 语句前缀的内容
suffix	表示在 trim 标签包裹的部分后面添加内容。注意：是在没有 prefixOverrides 和 suffixOverrides 属性的情况下
suffixOverrides	有 prefix 属性的情况下，suffixOverrides 属性表示去掉 SQL 语句后缀的内容

编写 SQL 语句的时候，通常喜欢写这样的 SQL 语句：

```
    <select id="findByName"  parameterType="String" resultType=
"com.ay.model.AyUser">
        SELECT * FROM ay_user WHERE 1 = 1
        <if test="name != null and name != ''">
            and name = #{name}
        </if>
    </select>
```

WHERE 1 = 1 这样的条件显然很奇怪，所以可以使用 where 标签优化上面的 SQL 语句，具体代码如下：

```
    <select id="findByName"  parameterType="String" resultType=
"com.ay.model.AyUser">
        SELECT * FROM ay_user
        <where>
            <if test="name != null and name != ''">
                and name = #{name}
            </if>
        </where>
    </select>
```

这样当 where 元素里面的条件成立的时候，才会加入 where 这个 SQL 关键字到组装的 SQL 里面，否则就不加入。

set 元素在执行 SQL 更新中会使用到，先来看一个传统代码的写法，具体代码如下：

```
<update id="update" parameterType="com.ay.model.AyUser">
    UPDATE ay_user SET
    name = #{name},
    password = #{password}
    WHERE id = #{id}
</update>
```

上面代码中没有使用 set 元素，可以对代码进行优化，具体优化后的代码如下：

```
<update id="update" parameterType="com.ay.model.AyUser">
    UPDATE ay_user
    <set>
        <if test="name != null and name != ''">
            name = #{name},
        </if>
        <if test="password != null and password != ''">
            password = #{password},
        </if>
    </set>
    WHERE id = #{id}
</update>
```

set 元素遇到逗号，它会把对应的逗号去掉，比如上面代码中的 password =

#{password}，不需要自己写判断语句去除逗号。

4.2.5 foreach 元素

foreach 元素是一个循环语句，作用是遍历集合，支持数组、List、Set 等。具体看下面的示例：

```
//根据 Id 集合查询用户列表
<select id="findByIds" resultType="com.ay.model.AyUser">
    SELECT * FROM ay_user
    WHERE id in
    <foreach item="item" index="index" collection="list"
           open="(" separator="," close=")">
       #{item}
    </foreach>
</select>
```

Foreach 的元素属性配置如表 4-10 所示。

表 4-10 foreach 元素的属性

属性名称	描 述
item	循环中当前的元素
index	配置的是当前元素在集合的位置下标
collection	接口传递进来的参数名称，可以是一个数组、List、Set 等集合
open	配置的是以什么符号将集合中的元素包装起来，如“（”
separator	各个元素的分隔符
close	配置的是以什么符号将集合中的元素包装起来，如“）”

4.2.6 bind 元素

bind 元素可以从 OGNL 表达式中创建一个变量并将其绑定到上下文。在进行模糊查询的时候，比如通过用户名称 name 查询用户，就常会用到 bind 元素，具体请看下面的示例：

```
<select id="findByNameAndPassword"
parameterType="String" resultType="com.ay.model.AyUser">
    <bind name="name_pattern" value="'%' + name + '%'"/>
    <bind name="password_pattern" value="'%' + password + '%'"/>
    SELECT * FROM ay_user
    <where>
       <if test="name != null and name != ''">
          and name LIKE #{name_pattern}
       </if>
       <if test="password != null and password != ''">
          and password LIKE #{password_pattern}
      </if>
```

```
    </where>
</select>
```

上述的 select 元素中，定义了多个 bind 元素，bind 元素的属性 value 的值："'%' + password + '%'" 会赋值给 name_pattern，然后就可以在 SQL 中直接使用 name_pattern 变量。

4.3 MyBatis 注解配置

4.3.1 MyBatis 常用注解

在前面的章节中，MySQL 的映射器、动态 SQL 语句等知识都是使用基于 XML 的配置方式，其实除了使用 XML 的配置方式，还可以使用基于注解的配置方式。MyBatis 提供了很多好用的注解，具体见表 4-11。

表 4-11　MyBatis 常用注解

属性名称	描　　述
@Select	映射查询 SQL 语句
@SelectProvider	Select 语句的动态 SQL 映射。允许指定一个类和一个方法在执行时返回运行的查询语句。有两个属性：type 和 method，type 属性是类的完全限定名，method 是该类中的那个方法名
@Delete	映射删除 SQL 语句
@DeleteProvider	Delete 语句的动态 SQL 映射。允许指定一个类和一个方法在执行时返回运行的删除语句。有两个属性：type 和 method，type 属性是类的完全限定名，method 是该类中的那个方法名
@Insert	映射插入 SQL 语句
@InsertProvider	Delete 语句的动态 SQL 映射。允许指定一个类和一个方法在执行时返回运行的插入语句。有两个属性：type 和 method，type 属性是类的完全限定名，method 是该类中的那个方法名
@Update	映射更新 SQL 语句
@UpdateProvider	Delete 语句的动态 SQL 映射。允许指定一个类和一个方法在执行时返回运行的更新语句。有两个属性：type 和 method，type 属性是类的完全限定名，method 是该类中的那个方法名
@Result	列和属性之间的单独结果映射。属性包括：id、column、property、javaType、jdbcType、type、Handler、one、many，其中 id 属性是一个布尔值，表示是否被用于主键映射
@Results	多个结果映射（Result）列表
@Options	提供配置选项的附加值，通常在映射语句上作为附加功能配置出现
@One	复杂类型的单独属性值映射

（续表）

属性名称	描　　述
@Many	复杂类型的集合属性映射
@Param	当映射器的方法需要多个参数时，注解可以被应用于映射器方法参数来给每个参数取个名字。否则，多参数默认将会以它们的顺序位置和 SQL 语句中的表达式进行映射

4.3.2　@Select 注解

@Select 注解与 XML 配置里的 select 标签相对应，@Results 注解与 resultMap 标签相对应，当在 AyUserDao 接口中使用注解的配置方式时，就不需要在 XML 里面配置，具体示例如下：

```
@Repository
public interface AyUserDao {
    //示例 1：查询所有的用户列表
@Select("SELECT * FROM ay_user")
    List<AyUser> findAll();
        //示例 2：查询所有的用户列表
        @Select("SELECT * FROM ay_user")
    @Results({
            @Result(id = true,column = "id",property = "id"),
            @Result(column = "name",property = "name"),
            @Result(column = "password",property = "password")
    })
    List<AyUser> findAll();
    //示例 3：通过 id 查询用户
    @Select("SELECT * FROM ay_user WHERE id = #{id}")
    AyUser findById(String id);
    //示例 4：通过用户名获取用户
    @Select("SELECT * FROM ay_user WHERE name = #{name}")
    List<AyUser> findByName(String name);

}
```

4.3.3　@Insert、@Update、@Delete 注解

@Insert、@Update、@Delete 注解与 XML 配置里的 insert、update、delete 标签相对应，具体示例如下：

```
@Repository
public interface AyUserDao {
//示例 1：插入用户数据
@Insert("INSERT INTO ay_user(name,password) VALUES(#{name}, #{password})")
@Options(useGeneratedKeys = true, keyProperty = "id")
```

```
    int insert(AyUser ayUser);
        //示例 2：更新用户数据
    @Update("UPDATE  ay_user SET name = #{name}, password = #{password} WHERE
id = #{id}")
        int update(AyUser ayUser);
        //示例 3：根据用户 id 删除用户
        @Delete("DELETE FROM ay_user WHERE id = #{id}")
        int delete(int id);
        //示例 4：根据用户名删除用户
        @Delete("DELETE FROM ay_user WHERE name = #{name}")
        int deleteByName(String name);

    }
```

@Options 注解提供配置选项的附加值，通常在映射语句上作为附加功能配置出现。

4.3.4　@Param 注解

当映射器的方法需要多个参数时，@Param 注解可以被应用于映射器方法参数来给每个参数取个名字。否则，多参数默认将会以它们的顺序位置和 SQL 语句中的表达式进行映射。具体示例如下：

```
    @Select("SELECT * FROM ay_user WHERE name = #{name} and password =
#{password}")
    List<AyUser> findByNameAndPassword(@Param("name") String name,
    @Param("password")String password);
```

除了使用@Param 注解映射参数之外，还有其他的方式来映射参数，只是不是很推荐，这里简单地总结一下其他的方法，大家可以做简单的对比。

1. Map的映射方式

AyUserDao 接口的方法定义如下：

```
    List<AyUser> findByNameAndPassword(Map<String, String > map);
```

对应的 XML 配置如下：

```
    <select id="findByNameAndPassword" parameterType="java.util.Map"
resultMap="userMap">
        SELECT * from ay_user a
        <where>
            <if test="name != null and name != ''">
                and name = #{name}
            </if>
            <if test="password != null and password != ''">
                and password = #{password}
            </if>
```

```
    </where>
</select>
```

服务层 AyUserServiceImpl 调用方式如下：

```
public List<AyUser> findByNameAndPassword(Map<String,String> map) {
        Map<String,String> map = new HashMap<String, String>();
        map.put("name","al");
        map.put("password","123");
        return ayUserDao.findByNameAndPassword(map);
}
```

2. 顺序映射方式

AyUserDao 接口的方法定义如下：

```
List<AyUser> findByNameAndPassword(String name,String password);
```

对应的 XML 配置如下：

```
<select id="findByNameAndPassword" parameterType="String"
resultMap="userMap">
    SELECT * from ay_user a
    WHERE 1 = 1 AND name = #{param1} AND password = #{param2}
</select>
```

顺序映射方式是通过参数的顺序进行映射的，name 参数对应#{param1}，#{password}参数对应 param2，当然这是新版的 MyBatis 提供的，旧版本的 MyBatis 是用#{0}、#{1}进行映射的。

4.4 MyBatis 关联映射

4.4.1 关联映射概述

之前的章节中，我们已介绍了使用 MyBatis 对数据库单表进行映射和执行增删改查操作，但是在现实的项目中进行数据库建模时，需要遵循数据库设计范式的要求，对现实中的业务模型进行拆分，封装到不同的数据表中，表与表之间存在着一对多或多对多的对应关系。进而，对数据库的增删改查操作的主体，也就从单表变成了多表。那么 MyBatis 中是如何实现这种多表关系的映射呢？这是接下来要学习的重点。

关联映射大致可以分为：一对一、一对多、多对多，下面将一一展开描述。

4.4.2 一对一

一对一关联关系在现实生活中很多，比如：一个人只能有一个身份证或者一个老家地址。首先，在数据库 springmvc-mybatis-book 中创建数据库表 ay_user、ay_user_address，具体代码如下：

```
-- ----------------------------
-- Table structure for ay_user
-- ----------------------------
DROP TABLE IF EXISTS 'ay_user';
CREATE TABLE 'ay_user' (
  'id' bigint(32) NOT NULL AUTO_INCREMENT,
  'name' varchar(10) DEFAULT NULL,
  'password' varchar(64) DEFAULT NULL,
  'age' int(10) DEFAULT NULL,
  'address_id' bigint(32) DEFAULT NULL,
  PRIMARY KEY ('id'),
  KEY 'FK_address_id' ('address_id'),
  CONSTRAINT 'FK_address_id' FOREIGN KEY ('address_id')
REFERENCES 'ay_user_address' ('id')
) ENGINE=InnoDB AUTO_INCREMENT=5 DEFAULT CHARSET=utf8;
-- ------------------------------------
-- Table structure for ay_user_address
-- ------------------------------------
DROP TABLE IF EXISTS 'ay_user_address';
CREATE TABLE 'ay_user_address' (
  'id' bigint(32) NOT NULL,
  'name' varchar(255) DEFAULT NULL,
  PRIMARY KEY ('id')
) ENGINE=InnoDB DEFAULT CHARSET=utf8;
```

表 ay_user、ay_user_address 对应的 Model 代码如下：

```
/**
 * 用户实体
 * @author Ay
 * @date  2018/04/02
 */
public class AyUser implements Serializable{
    private Integer id;
    private String name;
    private String password;
    private Integer age;
    //用户和地址一一对应，即一个用户只有一个老家地址
    private AyUserAddress ayUserAddress;
        //省略 set、get 方法
}

/**
 * 描述：用户地址实体
```

```
 * @author Ay
 * @create 2018/05/01
 **/
public class AyUserAddress  implements Serializable {

    private Integer id;
    private String name;

    //省略 set、get 方法
}
```

用户和老家地址是一对一关系，即一个用户只能有一个老家地址。在 AyUser 类中定义一个 ayUserAddress 属性，用来映射一对一的关联关系，表示一个人的老家地址。

AyUser 类和 AyUserAddress 类创建完成之后，创建对应的 DAO 接口 AyUserDao 和 AyUserAddressDao，具体代码如下：

```
@Repository
public interface AyUserDao {
    //根据 id 查询用户
    AyUser findById(String id);
}

@Repository
public interface AyUserAddressDao {
    //根据 id 查询用户地址
    AyUserAddress findById(Integer id);
}
```

DAO 接口 AyUserDao 和 AyUserAddressDao 创建完成之后，继续创建对应的 XML 配置文件 AyUserMapper.xml 和 AyUserAddressMapper.xml，AyUserAddressMapper.xml 具体代码如下：

```
<?xml version="1.0" encoding="UTF-8" ?>
<!DOCTYPE mapper PUBLIC "-//mybatis.org//DTD Mapper 3.0//EN"
        "http://mybatis.org/dtd/mybatis-3-mapper.dtd">
<mapper namespace="com.ay.dao.AyUserAddressDao">
    <select id="findById"
 parameterType="Integer" resultType="com.ay.model.AyUserAddress">
        SELECT * FROM ay_user_address WHERE id = #{id}
    </select>
</mapper>
```

AyUserMapper.xml 配置文件具体代码如下：

```
<?xml version="1.0" encoding="UTF-8" ?>
<!DOCTYPE mapper PUBLIC "-//mybatis.org//DTD Mapper 3.0//EN"
        "http://mybatis.org/dtd/mybatis-3-mapper.dtd">
<mapper namespace="com.ay.dao.AyUserDao">
```

```
    <resultMap id="userMap" type="com.ay.model.AyUser">
        <id property="id" column="id"/>
        <result property="name" column="name"/>
        <result property="password" column="password"/>
        <association property="ayUserAddress" column="address_id"
               select="com.ay.dao.AyUserAddressDao.findById"
  javaType="com.ay.model.AyUserAddress">
        </association>
    </resultMap>

    <select id="findById"  parameterType="String" resultMap="userMap">
      SELECT * FROM ay_user
      WHERE id = #{id}
    </select>

</mapper>
```

AyUserAddressMapper.xml 配置文件中通过<select>标签定义了一个通过 id 查询地址的简单查询。AyUserMapper.xml 配置文件中同样也配置了通过 id 查询用户的简单查询，查询返回的结果为 resultMap。resultMap 标签通过<association>元素映射一对一的关联关系，select 属性表示使用 column 属性的 address_id 的值作为参数执行 AyUserAddressDao 中定义的 findById 方法查询对应的地址数据，查询出的地址数据被封装到 property 属性的 ayUserAddress 对象当中。这样一对一的关联映射就完成了。

4.4.3　一对多

一对多关联关系在现实生活中也有很多，比如，一所学校可以包含多名学生，一名学生只属于一所学校，学校和学生就是一对多关系，而学生和学校就是多对一的关系。首先，在数据库 springmvc-mybatis-book 中创建数据库表 ay_student、ay_school，具体代码如下：

```
-- ------------------------------
-- Table structure for ay_student
-- ------------------------------
DROP TABLE IF EXISTS 'ay_student';
CREATE TABLE 'ay_student' (
  'id' bigint(32) NOT NULL,
  'name' varchar(255) DEFAULT NULL,
  'age' int(2) DEFAULT NULL,
  'school_id' bigint(32) DEFAULT NULL,
  PRIMARY KEY ('id')
) ENGINE=InnoDB DEFAULT CHARSET=utf8;

-- ----------------------------
-- Table structure for ay_school
-- ----------------------------
```

```
DROP TABLE IF EXISTS 'ay_school';
CREATE TABLE 'ay_school' (
  'id' bigint(32) NOT NULL,
  'name' varchar(255) DEFAULT NULL,
  PRIMARY KEY ('id')
) ENGINE=InnoDB DEFAULT CHARSET=utf8;
```

数据库表 ay_student 和 ay_school 创建完成之后，接着创建对应的实体类 AyStudent 和 AySchool，具体代码如下：

```
/**
 * 描述：学生实体
 * @author Ay
 * @create 2018/05/01
 **/
public class AyStudent implements Serializable {

    private Integer id;
    private String name;
    private Integer age;
    //一名学生只能在一所学校
    private AySchool aySchool;
    //省略 set、get 方法
}
```

一名学生只能在一所学校里，所以在 AyStudent 实体里维护 aySchool 属性，AySchool 实体类的代码如下：

```
/**
 * 描述：学校实体
 * @author Ay
 * @create 2018/05/01
 **/
public class AySchool implements Serializable {

    private Integer id;
    private String name;
    //一所学校有多名学生
    private List<AyStudent> students;

//省略 set、get 方法
}
```

一所学校里面有很多学生，所以在 AySchool 实体类中维护 students 的集合列表，该集合列表是一个 List 对象。实体类 AyStudent 和 AySchool 创建完成之后，继续创建 AyStudentDao 和 AySchoolDao 接口，具体代码如下：

```
/**
```

```
 * 描述：学生 DAO 接口
 * @author Ay
 * @create 2018/05/01
 **/
public interface AyStudentDao {
    List<AyStudent> findBySchoolId(Integer id);
}

/**
 * 描述：学校 DAO 接口
 * @author Ay
 * @create 2018/05/01
 **/
public interface AySchoolDao {

    AySchool findById(Integer id);
}
```

AyStudentDao 和 AySchoolDao 接口创建完成之后，创建对应的 XML 配置文件，AyStudentMapper.xml 具体代码如下：

```
<?xml version="1.0" encoding="UTF-8" ?>
<!DOCTYPE mapper PUBLIC "-//mybatis.org//DTD Mapper 3.0//EN"
      "http://mybatis.org/dtd/mybatis-3-mapper.dtd">
<mapper namespace="com.ay.dao.AyStudentDao">
    <resultMap id="studentMap" type="com.ay.model.AyStudent">
        <id property="id" column="id"/>
        <result property="name" column="name"/>
        <result property="age" column="age"/>
        <association property="aySchool" javaType="com.ay.model.AySchool">
            <id property="id" column="id"/>
            <result property="name" column="name"/>
        </association>
    </resultMap>
     <! -- 根据 id 查询学生，关联 ay_school 表 -->
    <select id="findById" parameterType="Integer" resultMap="studentMap">
        SELECT * FROM ay_student s , ay_school c
        WHERE s.school_id = c.id
        AND s.id = #{id}
    </select>
    <! -- 根据学校 id 查询学生 -->
<select id="findBySchoolId"
parameterType="Integer" resultType="com.ay.model.AyStudent">
        SELECT * FROM ay_student WHERE school_id = #{school_id}
    </select>
</mapper>
```

在 AyStudentMapper.xml 配置文件中，使用<select>标签，通过 id 查询学生实体，同时关联了 ay_school 表，查询结果返回到 studentMap 中。在 studentMap 中，使用< association >元素映射多对一的关联关系。因为<select id="findById" ..>的 SQL 语句是一条多表连接，在查询学生的同时，会把对应的学校查询出来。

AySchoolMapper.xml 具体代码如下：

```
<?xml version="1.0" encoding="UTF-8" ?>
<!DOCTYPE mapper PUBLIC "-//mybatis.org//DTD Mapper 3.0//EN"
        "http://mybatis.org/dtd/mybatis-3-mapper.dtd">
<mapper namespace="com.ay.dao.AySchoolDao">

    <resultMap id="schoolMap" type="com.ay.model.AySchool">
        <id property="id" column="id" />
        <result property="name" column="name"/>
        <collection property="students" javaType="ArrayList" column="id"
                    ofType="com.ay.model.AyStudent"
                    fetchType="lazy"
                     select="com.ay.dao.AyStudentDao.findBySchoolId">
            <id property="id" column="id"/>
            <result property="name" column="name"/>
            <result property="age" column="age"/>
        </collection>
    </resultMap>

    <!-- 根据 id 查询学校 -->
    <select id="findById" parameterType="Integer" resultMap="schoolMap">
        SELECT * FROM ay_school WHERE id = #{id}
    </select>
</mapper>
```

在 AySchoolMapper.xml 配置文件中，使用<select>标签，通过 id 查询学校实体，返回结果到 schoolMap 中。在 resultMap 中，使用<collection>元素映射一对多的关联关系，select 属性表示会使用 column 属性的 id 值作为参数执行 AyStudentDao 接口中的 findBySchoolId 查询该学校下所有的学生数据，查询出的数据将被封装到 property 表示的 students 对象中。

4.4.4 多对多

多对多关联关系在现实生活中也有很多，比如，用户和角色的关系，一个用户可以有多个角色，每个角色也可以有多个用户；学生和老师的关系，一名学生可以跟多位老师，每位老师也可以教多名学生等。首先，在数据库 springmvc-mybatis-book 中创建数据库表 ay_user、ay_role、ay_user_role_rel，具体代码如下：

```
-- ----------------------------
-- Table structure for ay_user
-- ----------------------------
```

```
DROP TABLE IF EXISTS 'ay_user';
CREATE TABLE 'ay_user' (
  'id' bigint(32) NOT NULL AUTO_INCREMENT,
  'name' varchar(10) DEFAULT NULL,
  'password' varchar(64) DEFAULT NULL,
  'age' int(10) DEFAULT NULL,
  'address_id' bigint(32) DEFAULT NULL,
  PRIMARY KEY ('id'),
  KEY 'FK_address_id' ('address_id'),
  CONSTRAINT 'FK_address_id' FOREIGN KEY ('address_id')
REFERENCES 'ay_user_address' ('id')
) ENGINE=InnoDB AUTO_INCREMENT=5 DEFAULT CHARSET=utf8;

-- ----------------------------
-- Table structure for ay_role
-- ----------------------------
DROP TABLE IF EXISTS 'ay_role';
CREATE TABLE 'ay_role' (
  'id' bigint(32) NOT NULL,
  'name' varchar(255) DEFAULT NULL,
  PRIMARY KEY ('id')
) ENGINE=InnoDB DEFAULT CHARSET=utf8;

-- ----------------------------
-- Table structure for ay_user_role_rel
-- ----------------------------
DROP TABLE IF EXISTS 'ay_user_role_rel';
CREATE TABLE 'ay_user_role_rel' (
  'id' bigint(32) NOT NULL,
  'user_id' bigint(32) DEFAULT NULL,
  'role_id' bigint(32) DEFAULT NULL,
  PRIMARY KEY ('id')
) ENGINE=InnoDB DEFAULT CHARSET=utf8;
```

数据库表 ay_user、ay_role、ay_user_role_rel 创建完成之后，创建对应的实体对象 AyUser、AyRole、AyUserRoleRel，具体代码如下：

```
/**
 * 用户实体
 * @author Ay
 * @date  2018/04/02
 */
public class AyUser implements Serializable{

   private Integer id;
   private String name;
```

```
    private String password;
    private Integer age;
    //用户与角色是多对多关系，一个用户有多个角色
    private List<AyRole> ayRoleList;
}

/**
 * 描述：角色实体
 *
 * @author Ay
 * @create 2018/05/01
 **/
public class AyRole implements Serializable {
    private Integer id;
    private String name;
    //角色与用户是多对多关系，一个角色对应多个用户
    private List<AyUser> ayUserList;
}

/**
 * 描述：用户角色关联实体
 * @author Ay
 * @create 2018/05/01
 **/
public class AyUserRoleRel implements Serializable {

    private Integer id;
    //用户 id
    private Integer userId;
    //角色 id
    private Integer roleId;
}
```

用户和角色是多对多关联关系，在 AyUser 实体对象中定义 List<AyRole> ayRoleList 角色属性，用来维护用户和角色的关系。同理，在 AyRole 实体对象中定义 List<AyUser> ayUserList 用户属性，用来维护角色与用户的关系。实体对象 AyUser、AyRole、AyUserRoleRel 创建完成之后，创建对应的 DAO 接口对象，具体代码如下：

```
@Repository
public interface AyUserDao {
     AyUser findById(String id);
}

@Repository
public interface AyRoleDao {
```

```
    AyRole findById(String id);
}
```

AyUserDao 和 AyRoleDao 接口创建完成之后，创建对应的 XML 配置文件 AyUserMapper.xml 和 AyRoleMapper.xml，AyUserMapper.xml 配置文件的具体代码如下：

```
<?xml version="1.0" encoding="UTF-8" ?>
<!DOCTYPE mapper PUBLIC "-//mybatis.org//DTD Mapper 3.0//EN"
        "http://mybatis.org/dtd/mybatis-3-mapper.dtd">
<mapper namespace="com.ay.dao.AyUserDao">

    <resultMap id="userMap" type="com.ay.model.AyUser">
        <id property="id" column="id"/>
        <result property="name" column="name"/>
        <result property="password" column="password"/>
        <collection property="ayRoleList" javaType="ArrayList" column="id"
                    ofType="com.ay.model.AyRole"
                    fetchType="lazy"
                    select="com.ay.dao.AyRoleDao.findByUserId">
            <id property="id" column="id"/>
            <result property="name" column="name"/>
        </collection>
    </resultMap>

    <select id="findById"  parameterType="String" resultMap="userMap">
      SELECT * FROM ay_user
      WHERE id = #{id}
    </select>

    <select id="findByRoleId" parameterType="Integer"
resultType="com.ay.model.AyUser">
        SELECT * FROM ay_user WHERE id in(
          select user_id from ay_user_role_rel where role_id = #{roleId}
        )
    </select>
</mapper>
```

AyRoleMapper.xml 配置文件的具体代码如下：

```
<?xml version="1.0" encoding="UTF-8" ?>
<!DOCTYPE mapper PUBLIC "-//mybatis.org//DTD Mapper 3.0//EN"
        "http://mybatis.org/dtd/mybatis-3-mapper.dtd">
<mapper namespace="com.ay.dao.AyRoleDao">

    <resultMap id="roleMap" type="com.ay.model.AyRole">
        <id property="id" column="id"/>
        <result property="name" column="name"/>
```

```
        <collection property="ayUserList" javaType="ArrayList" column="id"
                    ofType="com.ay.model.AyUser"
                    fetchType="lazy"
                    select="com.ay.dao.AyUserDao.findByRoleId">
            <id property="id" column="id"/>
            <result property="name" column="name"/>
            <result property="password" column="password"/>
        </collection>
    </resultMap>
    <select id="findById" parameterType="String" resultMap="roleMap">
        SELECT * FROM ay_role WHERE id = #{id}
    </select>
    <select id="findByUserId" parameterType="Integer"
resultType="com.ay.model.AyRole">
        SELECT * FROM ay_role WHERE id in(
          select role_id from ay_user_role_rel where user_id = #{userId}
        )
    </select>
</mapper>
```

在 AyUserMapper.xml 配置文件中，通过<select>标签查询用户信息，将返回的结果存放到 userMap 中。userMap 中定义<collection>元素映射一对多的关联关系，select 属性表示使用 column 属性的 id 值作为参数执行 AyRoleDao 接口中的 findByUserId 方法，查询出的数据将被封装到 property 表示的 ayRoleList 对象当中。

在 AyRoleMapper.xml 配置文件中，通过<select>标签查询角色信息，将返回的结果存放到 roleMap 中。roleMap 中定义<collection>元素映射一对多的关联关系，select 属性表示使用 column 属性的 id 值作为参数执行 AyUserDao 接口中的 findByRoleId 方法，查询出的数据将被封装到 property 表示的 ayUserList 对象当中，fetchType="lazy"表示使用懒加载的方式加载数据。

开启懒加载模式，需要在 mybatis-config.xml 配置文件中添加如下代码：

```
<?xml version="1.0" encoding="UTF-8" ?>
<!DOCTYPE configuration
      PUBLIC "-//mybatis.org//DTD Config 3.0//EN"
      "http://mybatis.org/dtd/mybatis-3-config.dtd">
<configuration>
   <!-- 全局配置参数，需要时再设置 -->
   <settings>
      //省略代码
      <!-- 通过日志记录显示 mybatis 的执行过程 -->
      <setting name="logImpl" value="log4j" />
      <!-- lazyLoadingEnabled 设置为懒加载 -->
      <setting name="lazyLoadingEnabled" value="true"/>
      <!-- aggressiveLazyLoading 主动加载为 false -->
```

```
        <setting name="aggressiveLazyLoading" value="false"/>
    </settings>
</configuration>
```

4.5　思考与练习

1. 简述 MyBatis 提供了哪些映射器元素。
2. 简述 select 元素有哪些元素，id 元素的作用是什么？
3. 简述 select、insert、update、delete 元素的作用。
4. SQL 语句中$和#区别是什么？
5. 简述常用动态 SQL 元素有哪些？
6. 简述 trim 元素的作用。
7. 简述 foreach 元素作用，foreach 元素的属性有哪些？
8. @Select 注解对应映射器的哪个元素？
9. 使用@Select、@Insert、@Update 注解实现简单的查询。
10. MyBatis 关联映射大致可以分为几类？
11. 动手实现一对一关联映射查询。

第 5 章

Spring MVC 常用注解

本章将介绍 Spring MVC 常用注解，包括请求映射注解和参数绑定注解以及 Spring MVC 信息转换的原理。

5.1 请求映射注解

5.1.1 @Controller 注解

在 Spring MVC 中，控制器 Controller 负责处理由 DispatcherServlet 分发的请求，它把用户请求的数据经过业务处理层处理之后封装成一个 Model，然后再把该 Model 返回给对应的 View 视图层进行展示。在 Spring MVC 中提供了一个非常简便的定义 Controller 的方法，你无须继承特定的类或实现特定的接口，只需使用@Controller 标记一个类是 Controller，然后使用@RequestMapping 和@RequestParam 等一些注解用以定义 URL 请求和 Controller 方法之间的映射，这样的 Controller 就能被外界访问到。此外 Controller 不会直接依赖于 HttpServletRequest 和 HttpServletResponse 等 HttpServlet 对象，它们可以通过 Controller 的方法参数灵活地获取到。为了对 Controller 有一个初步的印象，以下先定义一个简单的 Controller，具体源码如下：

```
/**
 * 用户控制层
 *@author Ay
 * @date 2018/04/02
 */
```

```
@Controller
@RequestMapping(value = "/user")
public class AyUserController {

    @GetMapping("/hello")
    public String hello(Model model){
        model.addAttribute("message", "hello ay");
        return "hello";
    }
}
```

上述代码定义了一个 AyUserController 控制层，使用@Controller 注解进行表示，使用@GetMapping 注解来映射一个请求，value="/hello"。为了保证 Spring 能够找到控制层，需要额外进行配置：在 applicationContext.xml 配置文件中配置<context:component-scan />元素，如果已经配置可以略过，具体代码如下：

```
<context:component-scan base-package="com.ay. controller "/>
```

<context:component-scan/>元素的功能是，启动包扫描功能，使加有@Controller、@Service、@Respository、@Component 等注解的类能够成为 Spring 的 bean。base-package 属性指定了需要扫描的类包，类包及其包含的子包中的所有类都会被处理。

此外，还需要在 web.xml 文件中配置 Spring MVC 的前端控制器 DispatcherServlet，以及在 spring-mvc.xml 配置文件中配置 InternalResourceViewResolver 视图解析器，具体配置在第 2 章已详细介绍，这里不再赘述。

启动项目 springmvc-mybatis-book，在浏览器中输入 URL 路径：http://localhost:8080/user/hello，会看到如图 5-1 所示的图片，表示 Spring MVC 访问成功。

图 5-1　AyUserController 控制层成功访问界面

5.1.2　@RequestMapping 注解

@RequestMapping 是 Spring Web 应用程序中最常用的注解之一，这个注解会将 HTTP 请求映射到 MVC 和 RES 控制器的处理方法上。@RequestMapping 注解可以在控制器类的级别或方法的级别上使用，在类级别上的注解会将一个特定请求或者请求模式映射到一个控制器之上，之后你还可以另外添加方法级别的注解来进一步指定到处理方法的映射关系。先来看一个具体示例：

```
@Controller
@RequestMapping(value = "/user")
```

```
public class AyUserController {

    @RequestMapping("/hello")
    public String hello(Model model){
        model.addAttribute("message", "hello ay");
        return "hello";
    }
}
```

上述代码中，当要访问 AyUserController 类中的 hello 方法时，可在浏览器中输入请求 URL：http://localhost:8080/user/hello。

@RequestMapping 注解除了 value 属性外，还有很多其他的属性，具体如表 5-1 所示。

表 5-1 @RequestMapping 的常用属性

属性名称	类　　型	是否必填	描　　述
value	String[]	否	指定请求的实际地址
name	String	否	给映射地址指定一个别名
method	RequestMethod[]	否	指定请求的 method 类型，包括 GET、POST、PUT、DELETE 等
consumer	String[]	否	指定处理请求的提交内容类型（Content-Type），例如 application/ json, text/html
produces	String[]	否	指定返回的内容类型，返回的内容类型必须是 request 请求头中的（Accept）类型中包含该指定类型
params	String[]	否	指定 request 中必须包含某些参数值时才能让该方法处理
headers	String[]	否	指定 request 中必须包含某些指定的 header 值时才能让该方法处理请求

我们来看几个示例，具体代码如下：

```
@Controller
@RequestMapping(value = "/user")
public class AyUserController {

    @RequestMapping(value = {
            "",
            "/page",
            "page*",
            "view/*,**/msg"
    })
    public String hello(Model model){
        model.addAttribute("message", "hello ay");
        return "hello";
```

```
    }
}
```

如上代码可知，可以将多个请求映射到一个方法上，只需要添加一个带有请求路径值列表的@RequestMapping 注解就可以了。

在浏览器中访问如下 URL，都会定位到 hello 方法上：

```
localhost:8080/user
localhost:8080/user/
localhost:8080/user/page
localhost:8080/user/pageabc
localhost:8080/user/view/
localhost:8080/user/view/view
```

Spring MVC 的@RequestMapping 注解能够处理 HTTP 请求的方法，比如 GET、PUT、POST、DELETE 以及 PATCH，所有的请求默认是 HTTP GET 类型的。为了将一个请求映射到一个特定的 HTTP 方法上，需要在@RequestMapping 中使用 method 属性来声明 HTTP 请求所使用的方法类型，如下所示：

```
@Controller
@RequestMapping(value = "/user")
public class AyUserController {

    @RequestMapping(method = RequestMethod.GET)
    String get() {
        return "Hello from get";
    }
    @RequestMapping(method = RequestMethod.DELETE)
    String delete() {
        return "Hello from delete";
    }
    @RequestMapping(method = RequestMethod.POST)
    String post() {
        return "Hello from post";
    }
    @RequestMapping(method = RequestMethod.PUT)
    String put() {
        return "Hello from put";
    }
    @RequestMapping(method = RequestMethod.PATCH)
    String patch() {
        return "Hello from patch";
    }
}
```

在上述代码中，@RequestMapping 注解中的 method 元素声明了 HTTP 请求方法的类型。所有的处理方法会处理同一个 URL（/user）进来的请求，但具体要看指定的 HTTP 方法类型来决定用哪个方法处理。例如，一个 POST 类型的请求/user 会交给 post()方法来处理，而一个 DELETE 类型的请求/user 则会由 delete()方法来处理。

可以使用@RequestMapping 注解的 produces 和 consumes 属性来指定处理请求的提交内容类型（Content-type）和返回的内容类型，返回的类型必须是 request 请求头（Accept）中所包含的类型。具体可以看下面的实例：

```
@Controller
@RequestMapping(value = "/user")
public class AyUserController {

    @RequestMapping(value = "/produces", produces = {
            "application/json"
    })
    @ResponseBody
    String getProduces() {
        return "Produces attribute";
    }

    @RequestMapping(value = "/consumes", consumes = {
            "application/json",
            "application/xml"
    })
    String getConsumes() {
        return "Consumes attribute";
    }
}
```

上述代码中，getProduces()方法会返回 JSON 类型的响应（application/json），getConsumes() 处理方法可以同时处理请求中的 JSON 和 XML 内容。

@RequestMapping 注解提供了一个 header 元素来根据请求中的消息头内容缩小请求映射的范围。该属性指定 request 中必须包含某些指定的 header 值，才能让该方法处理请求。具体示例如下：

```
@Controller
@RequestMapping(value = "/user")
public class AyUserController {

    @RequestMapping(value = "/head", headers = {
            "content-type=text/plain"
    })
    String post() {
        return "Mapping applied along with headers";
    }
}
```

上述代码中，@RequestMapping 注解的 headers 属性将映射范围缩小到 post()方法。post()方法只会处理/user/head 请求路径，并且 content-type 被指定为 text/plain 的请求。当然，还可以指定多个消息头。具体代码如下：

```
@Controller
@RequestMapping(value = "/user")
public class AyUserController {

    @RequestMapping(value = "/head", headers = {
            "content-type=text/plain",
            "content-type=text/html"
    }) String post() {
        return "Mapping applied along with headers";
    }
}
```

上述代码中，post()方法能同时接受请求类型为 text/plain 和 text/html 的请求。

@RequestMapping 可以通过 params 属性帮助进一步缩小请求映射的定位范围。使用 params 属性，可以让多个处理方法处理同一个 URL 的请求，而这些请求的参数是不一样的。可以用 myParams = myValue 这种格式来定义参数，也可以使用通配符来指定特定的参数值在请求中是不支持的。具体代码如下：

```
@Controller
@RequestMapping(value = "/user")
public class AyUserController {

    @RequestMapping(value = "/fetch", params = {
            "personId=10"
    })
    public String getParams(@RequestParam("personId") String id) {
        return "Fetched parameter using params attribute = " + id;
    }
    @RequestMapping(value = "/fetch", params = {
            "personId=20"
    })
    public String getParamsDifferent(@RequestParam("personId") String id) {
        return "Fetched parameter using params attribute = " + id;
    }
}
```

上述代码中，getParams()和 getParamsDifferent()两个方法都能处理相同的一个 URL（/user/fetch），但会根据 params 属性的配置不同而决定执行哪一个方法。例如，当 URL 是 /user /fetch?id=10, getParams()方法会执行，因为 id 的值为 10。对于 user/fetch?personId=20 请求 URL，getParamsDifferent()方法会执行，因为 id 值为 20。

5.1.3 @GetMapping 和@PostMapping 注解

@GetMapping 是一个组合注解，是@RequestMapping(method = RequestMethod.GET)的缩写。该注解将 HTTP GET 请求映射到特定的处理方法上。类似的@PostMapping 注解是@RequestMapping(method = RequestMethod.POST) 的缩写。@PutMapping 注解是@RequestMapping(method = RequestMethod.PUT) 的缩写。@DeleteMapping 注解是@RequestMapping(method = RequestMethod.DELETE) 的缩写。@PatchMapping 注解是@RequestMapping(method = RequestMethod.PATCH)的缩写。具体代码如下：

```
@Controller
@RequestMapping(value = "/user")
public class AyUserController {

    @GetMapping("/findById/{id}")
    public String findById(@PathVariable String id) {
        // ...
        return "";
    }

    @PostMapping(path = "/add")
    public String add(@RequestBody AyUser ayUser) {
        // ...
        return "";
    }
}
```

在浏览器中输入请求 URL：http://localhost:8080/user/findById/1 时，请求中的 id=1 会将值赋给 findById 方法中的 id 变量。在浏览器中输入 URL：http://localhost:8080/user/add 时，同时该请求为 POST 请求时，会调用 add 方法添加用户。

5.1.4 Model 和 ModelMap

Spring MVC 内部使用一个 org.springframework.ui.Model 接口存储数据模型，它的功能类似于 java.uitl.Map。org.springframework.ui.ModelMap 实现 Map 接口。Spring MVC 在调用方法前会创建一个隐含的数据模型，作为模型数据的存储容器。如果处理方法入参为 Map 或者 Model 类型，Spring MVC 会将隐含模型的引用传递给 Map 或者 Model，这样就可以在 Map 或者 Model 模型中访问所有的数据，可以向模型中添加新的属性数据。具体代码如下：

```
@Controller
@RequestMapping("/user")
public class AyUserController {

    @ModelAttribute
    public void redirectTest(Model model){
```

```
        model.addAttribute("name","ay");
    }

    @RequestMapping("hello")
    public String hello(Model model, ModelMap modelMap, Map map){
        return "hello";
    }
}
```

在浏览器中输入请求 URL：http://localhost:8080/user/hello 时，由于 redirectTest 方法添加@ModelAttribute 注解，故 redirectTest 方法优先执行。在 redirectTest 方法中设置 name 属性的值为“ay”，在 hello 方法中 Model、ModelMap 和 Map map 对象都可以获取到该 name 的属性值。虽然 redirectTest 方法（Model model, ModelMap modelMap, Map map）定义了三个不同的类型参数，但三者是同一个对象。

5.1.5 ModelAndView

当控制器处理完请求时，通常会将包含视图信息和模型数据信息的 ModelAndView 对象返回，这样 Spring MVC 将使用包含的视图对模型数据进行渲染。具体代码如下：

```
@Controller
@RequestMapping("/user")
public class AyUserController {

    @RequestMapping("hello")
    public ModelAndView hello(){
        ModelAndView mv = new ModelAndView();
        mv.addObject("name","ay");
        mv.setViewName("hello");
        return mv;
    }
}
```

在处理方法中可以使用 ModelAndView 对象的 addObject 方法添加属性和值，可以使用 setViewName 方法设置视图名称。上述代码等价于：

```
@Controller
@RequestMapping("/user")
public class AyUserController {

    @RequestMapping("hello")
    public String hello(Model model){
        model.addAttribute("name", "ay");
        return "hello";
    }
}
```

5.1.6 请求方法可出现参数和可返回类型

请求方法可出现的参数中，除了 Model 和 ModelMap 对象外，还可以出现其他的对象。具体如表 5-2 所示。

表 5-2 请求方法可出现的参数

属性名称	描　　述
ServletRequest ServletResponse HttpServletRequest HttpServletResponse	请求或响应对象（Servlet API）
HttpSession	HttpSession 类型的会话对象（Servlet API）
java.util.Locale	当前请求的地区信息 java.util.Locale，由已配置的最相关的地区解析器解析得到。在 MVC 的环境下，就是应用中配置的 LocaleResolver 或 LocaleContextResolver
java.util.TimeZone	与当前请求绑定的时区信息 java.util.TimeZone（java 6 以上的版本）/java.time.ZoneId（java 8），由 LocaleContextResolver 解析得到
org.springframework. http.HttpMethod	获取 HTTP 请求方法
java.security.Principal	包装了当前被认证的用户信息
HttpEntity<?>	提供了对 HTTP 请求头和请求内容的存取。请求流是通过 HttpMessageConverter 被转换成 entity 对象的
org.springframework.web.servlet. mvc.support.RedirectAttributes	用以指定重定向下使用的属性集以及添加 flash 属性（暂存在服务端的属性，它们会在下次重定向请求的范围中有效）
org.springframework. validation.Errors 或 org.springframework. validation.BindingResult	验证结果对象，用于存储前面的命令或表单对象的验证结果（紧接其前的第一个方法参数）
org.springframework. web.bind.support.SessionStatus	用以标记当前的表单处理已结束，这将触发一些清理操作：@SessionAttributes 在类级别标注的属性将被移除
org.springframework.web .util.UriComponentsBuilder	用于构造当前请求 URL 相关的信息，比如主机名、端口号、资源类型（scheme）、上下文路径、servlet 映射中的相对部分（literal part）等

下面来看几个具体示例。

如果需要访问 HttpServletRequest 对象，可以将其添加到方法中作为参数，Spring 会将对象正确的传递方法。

```
@RequestMapping("hello")
public ModelAndView hello(HttpServletRequest request){
    ModelAndView mv = new ModelAndView();
    mv.addObject("name","ay");
```

```
    mv.setViewName("hello");
    return mv;
}
```

如果需要访问 HttpMethod 对象，可以将其添加到方法中作为参数，Spring 会将对象正确的传递给方法。

```
@RequestMapping("hello")
public ModelAndView hello(HttpMethod method){
    ModelAndView mv = new ModelAndView();
    mv.addObject("name","ay");
    mv.setViewName("hello");
    return mv;
}
```

请求方法可返回的参数中，除了 ModelAndView 对象外，还可以出现其他的对象。下面列举几个常用的返回值，具体如表 5-3 所示。

表 5-3　请求方法可返回的参数

属性名称	描　　述
Model、ModelMap、Map	模型对象
String	返回值为 String 类型，有三种用处： （1）表示返回逻辑视图名 真正视图（JSP 路径）=前缀+逻辑视图名+后缀，例如： return "index" （2）redirect 重定向 redirect 重定向特点：浏览器地址栏中的 URL 会变化。修改提交的 request 数据无法传到重定向的地址，因为重定向后需重新进行 request（request 无法共享），例如： return "redirect:xxx.action"　重定向到另一个 Action 请求中 （3）forward 页面转发 通过 forward 进行页面转发，浏览器地址栏 URL 不变，request 可以共享。例如： return "forward:xxx.action"
void	如果处理器方法中已经对 response 响应数据进行了处理（比如在方法参数中定义一个 ServletResponse 或 HttpServletResponse 类型的参数并直接向其响应体中写东西），那么方法可以返回 void
HttpEntity<?>或 ResponseEntity<?>	提供对 Servlet HTTP 响应头和响应内容的存储，对象会被 HttpMessageConverters 转换成响应流

5.2 参数绑定注解

5.2.1 @RequstParam 注解

@RequstParam 注解用于将制定的请求参数赋值给方法中的形参。@RequstParam 注解可以使用的属性如表 5-4 所示。

表 5-4 @RequstParam 注解的常用属性

属性名称	类 型	是否必填	描 述
name	String	否	指定请求头绑定的名称
value	String	否	name 属性的别名
required	Boolean	否	指定参数是否必须绑定
defaultValue	String	否	请求没有传递参数而使用的默认值

@RequstParam 注解的使用示例如下：

```
@Controller
@RequestMapping(value = "/user")
public class AyUserController {

    @RequestMapping("findById")
    public String findById(@RequestParam(value="id") String id){
        AyUser ayUser = ayUserService.findById(id);
        return "success";
    }
}
```

当浏览器中输入 URL 请求 http://localhost:8080/user/findById?id=1 时，请求中的 id=1 会将值赋给 findById 方法中的 id 变量。@RequstParam 注解可以使用 required 属性指定参数是否必须传值。如果参数没有接收到任何值，可以使用 defaultValue 指定参数的默认值。具体代码如下：

```
@Controller
@RequestMapping(value = "/user")
public class AyUserController {

    @RequestMapping("/findByNameAndPassword")
    public String findByNameAndPassword(
                @RequestParam(value="name") String name,
                @RequestParam(value="password",required = false,
defaultValue = "123") String password){
        System.out.println("name=" + name);
```

```
        System.out.println("password" + password);
        return "success";
    }
}
```

当浏览器中输入 URL 请求：http://localhost:8080/user/findByNameAndPassword?name=ay 时，name 参数被赋值为 ay，由于 password 没传任何值，故默认值为 123。

6.2.2 @PathVariable 注解

@PathVariable 注解可以将 URL 中动态参数绑定到控制器处理方法的入参中。@PathVariable 注解只有一个 value 属性，类型为 String，表示绑定的名称，如果缺省则默认绑定同名参数。具体代码如下：

```
@Controller
@RequestMapping(value = "/user")
public class AyUserController {

    @RequestMapping("/owners/{ownerId}/pets/{petId}")
    public String findPet(@PathVariable Long ownerId, @PathVariable Long
petId) {
        // ...
        return "";
    }
}
```

当在浏览器中输入请求 URL：http://localhost:8080/user/owners/123/pets/456 时，则自动将动态参数{ownerId}和{petId}的值 123 和 456 绑定到@PathVariable 注解的同名参数上，即 ownerId = 123，petId = 456。

URL 中的动态参数除了可以绑定到方法上，还可以绑定到类上，具体代码如下：

```
@Controller
@RequestMapping(value = "/owners/{ownerId}")
public class AyUserController {

    @RequestMapping("/pets/{petId}")
    public String findPet(@PathVariable Long ownerId, @PathVariable Long
petId) {
        // ...
        return "";
    }
}
```

当在浏览器中输入请求 URL：http://localhost:8080/owners/123/pets/456 时，则自动将类上的动态参数{ownerId}和方法上的动态参数{petId}的值绑定到@PathVariable 注解的同名参数上，即 ownerId = 123，petId = 456。

5.2.3 @RequestHeader 注解

@RequestHeader 注解可以将请求的头信息数据映射到处理的方法参数上，@RequestHeader 注解的属性如表 5-5 所示。

表 5-5 @RequestHeader 注解的常用属性

属性名称	类　型	是否必填	描　述
Name	String	否	指定请求头绑定的名称
Value	String	否	name 属性的别名
required	Boolean	否	指定参数是否必须绑定
defaultValue	String	否	请求没有传递参数而使用的默认值

一般请求头信息如下所示：

```
Host                localhost:8080
Accept              text/html,application/xhtml+xml,application/xml;q=0.9
Accept-Language     fr,en-gb;q=0.7,en;q=0.3
Accept-Encoding     gzip,deflate
Accept-Charset      ISO-8859-1,utf-8;q=0.7,*;q=0.7
Keep-Alive          300
```

下面通过@RequestHeader 注解获取 Accept-Encoding 和 Keep-Alive 信息，具体代码如下：

```
@Controller
@RequestMapping(value = "/user")
public class AyUserController {

    @RequestMapping("/requestHeader")
    public String handle(
            @RequestHeader("Accept-Encoding") String[] encoding,
            @RequestHeader("Accept") String[] accept) {
        //...
        return "";
    }
}
```

当在浏览器中输入请求 URL：http://localhost:8080/user/requestHeader 时，则自动将请求头“Accept-Encoding”和“Accept”的值赋到 encoding 和 accept 变量上，由于请求头“Accept-Encoding”和“Accept”的数据类型是数组，所有定义 encoding 和 accept 变量为 String[]类型。

5.2.4 @CookieValue 注解

@CookieValue 注解用于将请求的 Cookie 信息映射到处理的方法参数上，@CookieValue

注解的属性如表5-6所示。

表5-6 @CookieValue注解的常用属性

属性名称	类　型	是否必填	描　述
Name	String	否	指定请求头绑定的名称
Value	String	否	name属性的别名
required	Boolean	否	指定参数是否必须绑定
defaultValue	String	否	请求没有传递参数而使用的默认值

关于@CookieValue注解的使用请看下面的示例：

```
@Controller
@RequestMapping(value = "/user")
public class AyUserController {

    @RequestMapping("/cookieValue")
    public String handle(@CookieValue("JSESSIONID") String cookie) {
        //...
        return "";
    }
}
```

当在浏览器中输入请求URL：http://localhost:8080/user/cookieValue时，则自动将JSESSIONID值设置到Cookie参数上。

5.2.5 @ModelAttribute 注解

@ModelAttribute注解主要是将请求参数绑定到Model对象上。@ModelAttribute注解只有一个value属性，类型为String，表示绑定的属性名称。当Controller类中有任意一个方法被@ModelAttribute注解标记，页面请求只要进入这个控制器，不管请求哪个方法，均会先执行被@ModelAttribute标记的方法，所以可以用@ModelAttribute注解的方法做一些初始化操作。当同一个Controller类中有多个方法被@ModelAttribute注解标记，所有被@ModelAttribute标记的方法均会被执行，按先后顺序执行，然后再进入请求的方法。具体代码如下：

```
@Controller
@RequestMapping(value = "/user")
public class AyUserController {

    @ModelAttribute
    public void init(){
        System.out.println("init ...");
    }

    @ModelAttribute
```

```
    public void init02(){
        System.out.println("init 02 ...");
    }

    @GetMapping("/findById/{id}")
    public String findById(@PathVariable String id) {
        // ...
        return "";
    }

    @ModelAttribute
    public void init03(){
        System.out.println("init 03 ...");
    }
}
```

当在浏览器输入访问 URL：http://localhost:8080/user/findById/1 时，便可以在控制台看到打印信息：

```
init ...
init 02 ...
init 03 ...
```

@ModelAttribute 注解有很多的额外使用方式，下面逐一进行介绍。

1. @ModelAtterbute方法无返回值的情况

```
@Controller
@RequestMapping(value = "/user")
public class AyUserController {

    @ModelAttribute
    public void init(Model model){
        AyUser ayUser = new AyUser();
        ayUser.setId(1);
        ayUser.setName("ay");
        model.addAttribute("user", ayUser);
    }

    @GetMapping("/hello")
    public String hello(){
        return "hello";
    }
}
```

上述代码中@ModelAttribute 注解标记的 init 方法无任何返回值，在 init 方法中创建一个用户对象 AyUser 并设置 id 和 name 的值，最后调用 Model 对象的 addAttribute 方法设置到

Model 对象中。对应前端 src\main\webapp\WEB-INF\views\hello.jsp 页面代码如下所示：

```
<%@page language="java" contentType="text/html; charset=UTF-8"
      pageEncoding="UTF-8" isELIgnored="false"%>
<!DOCTYPE HTML>
<html>
<head>
   <title>Getting Started: Serving Web Content</title>
   <meta http-equiv="Content-Type" content="text/html; charset=UTF-8" />
</head>
<body>
hello, ${user.name}
</body>
</html>
```

当在浏览器中输入请求 URL：http://localhost:8080/user/hello 时，浏览器显示“hello，ay”。

从执行结果可以看出，当执行请求时，首先访问 init 方法，然后再访问 hello 方法，并且是同一个请求。因为 model 模型数据的作用域与 request 相同，所以可以用@ModelAttribute 注解直接标记在方法上对实际要访问的方法进行一些初始化操作。

2. @ModelAttribute标记方法有返回值

```
@Controller
@RequestMapping(value = "/user")
public class AyUserController {

   @ModelAttribute("name")
   public String init(@RequestParam(value = "name", required = false)
String name){
      return name;
   }

   @GetMapping("/hello")
   public String hello(){
      return "hello";
   }
}
```

上述代码中，@ModelAttribute 注解标注的 init 方法带有返回值，@ModelAttribute("name")的 value 属性值为 key，而 init 方法值为 value，类似于：

```
model.addAttribute("name",name);
```

对应前端 src\main\webapp\WEB-INF\views\hello.jsp 页面的代码如下：

```
<%@page language="java" contentType="text/html; charset=UTF-8"
      pageEncoding="UTF-8" isELIgnored="false"%>
<!DOCTYPE HTML>
```

```
<html>
<head>
    <title>Getting Started: Serving Web Content</title>
    <meta http-equiv="Content-Type" content="text/html; charset=UTF-8" />
</head>
<body>
hello, ${name}
</body>
</html>
```

当在浏览器中输入请求 URL：http://localhost:8080/user/hello 时，浏览器显示“hello，ay”。

3. @ModelAttribute注解和@RequestMapping注解

@ModelAttribute 注解和@RequestMapping 注解同时标记在一个方法上。

```
@Controller
public class AyUserController {

    @ModelAttribute("name")
    @RequestMapping(value = "/hello")
    public String hello(){
        return "ay";
    }
}
```

上述代码中，@ModelAttribute 注解和@RequestMapping 注解同时标记在 hello 方法上，同时我们把 AyUserController 类上的@RequestMapping 注解去掉。此时，hello 方法的返回值“ay”并不是视图名称，而是 model 属性的值，视图名称是@RequestMapping 的 value 值“hello”。Model 的属性名称由@ModelAttribute 的 value 值指定，相当于在 request 中封装了 name（key）= ay（value）。对应前端 src\main\webapp\WEB-INF\views\hello.jsp 页面的代码如下：

```
<%@page language="java" contentType="text/html; charset=UTF-8"
        pageEncoding="UTF-8" isELIgnored="false"%>
<!DOCTYPE HTML>
<html>
<head>
    <title>Getting Started: Serving Web Content</title>
    <meta http-equiv="Content-Type" content="text/html; charset=UTF-8" />
</head>
<body>
hello, ${name}
</body>
</html>
```

当在浏览器中输入请求 URL：http://localhost:8080/hello 时，浏览器显示“hello，ay”。

4. 使用@ModelAttribute注解方法的参数

```
@Controller
@RequestMapping("/user")
public class AyUserController {

    @ModelAttribute("ayUser")
    public AyUser init(@RequestParam("id") Integer id,
                       @RequestParam("name") String name){
        AyUser ayUser = new AyUser();
        ayUser.setId(id);
        ayUser.setName(name);
        return ayUser;
    }

    @RequestMapping(value="hello")
    public String hello(@ModelAttribute("ayUser") AyUser ayUser){

        return "hello";
    }
}
```

当在浏览器中输入请求 URL：http://localhost:8080/user/hello?id=1&name=ay 时，会优先执行 init 方法，把 id 和 name 的值赋值给 init 方法绑定的参数，同时构造 AyUser 对象返回。这里 model 的属性名称就是@ModelAttribute("ayUser")的 value 值“ayUser”，model 的属性值就是 init 方法的返回值。hello 方法的参数 AyUser 使用了注解@ModelAttribute("ayUser")，表示参数 ayUser 的值就是 init 方法中的 model 属性。

对应前端 src\main\webapp\WEB-INF\views\hello.jsp 页面的代码如下：

```
<%@page language="java" contentType="text/html; charset=UTF-8"
        pageEncoding="UTF-8" isELIgnored="false"%>
<!DOCTYPE HTML>
<html>
<head>
    <title>Getting Started: Serving Web Content</title>
    <meta http-equiv="Content-Type" content="text/html; charset=UTF-8" />
</head>
<body>
hello, ${ayUser.name}
</body>
</html>
```

5.2.6 @SessionAttribute 和@SessionAttributes 注解

@ModelAttribute 注解作用在方法或者方法的参数上，表示将被注解的方法的返回值或者被注解的参数作为 Model 的属性加入到 Model 中，Spring 框架会将 Model 传递给前端。

Model 的生命周期只存在于 HTTP 请求的处理过程中，请求处理完成后，Model 就销毁了。如果想让参数在多个请求间共享，那么需要用到@SessionAttributes 注解。@SessionAttributes 注解只能声明在类上，不能声明在方法上。

@SessionAttributes 注解常用的属性如表 5-7 所示。

表 5-7　@RequestHeader 注解的常用属性

属性名称	类　　型	是否必填	描　　述
names	String[]	否	需要存储到 session 中数据的名称
value	String	否	name 属性的别名
types	Class<?>[]	否	根据指定参数的类型将模型中对应类型的参数存储到 session 中

下面看具体实例，具体代码如下：

```
@Controller
@SessionAttributes("ayUser")
@RequestMapping("/user")
public class AyUserController {

    @RequestMapping("redirect")
    public String redirectTest(Model model){
        AyUser ayUser = new AyUser();
        ayUser.setName("ay");
        model.addAttribute("ayUser",ayUser);
        return "redirect:hello";
    }

    @RequestMapping("hello")
    public String hello(ModelMap modelMap){
        AyUser ayUser = (AyUser) modelMap.get("ayUser");
        System.out.println(ayUser.getName());
        return "hello";
    }
}
```

在浏览器中输入请求 URL：http://localhost:8080/user/redirect 时，由于 AyUserController 类上添加了注解@SessionAttributes("ayUser")，方法 redirectTest 在执行过程中，将 AyUser 对象存放到 Model 对象的同时也会把对象存放到 HttpSession 作用域中。redirectTest 方法执行完成之后，会重定向到 hello 方法，在 hello 方法中 HttpSession 对象会将@SessionAttributes 注解的属性写入到新的 Model 中，所以可以通过 ModelMap 获取的 AyUser 对象打印信息。

除了使用 ModelMap 来接收 HttpSession 对象中的值外，还可以使用@SessionAttribute 注解，具体代码如下：

```
@Controller
@SessionAttributes("ayUser")
```

```
@RequestMapping("/user")
public class AyUserController {

    @RequestMapping("redirect")
    public String redirectTest(Model model){
        AyUser ayUser = new AyUser();
        ayUser.setName("ay");
        model.addAttribute("ayUser",ayUser);
        return "redirect:hello";
    }

    @RequestMapping("hello")
    public String hello(@SessionAttribute AyUser ayUser){
        System.out.println(ayUser.getName());
        return "hello";
    }
}
```

上述代码中，在 hello 方法中使用@SessionAttribute 注解来获取@SessionAttributes("ayUser")注解中的“ayUser”对象并赋值给 AyUser 对象，这样就可以很方便地在 hello 方法中使用 AyUser 对象。

如果想删除 HttpSession 对象中共享的属性，可以使用 SessionStatus.setComplete()只删除通过@SessionAttribute 保存到 HttpSession 中的属性。具体代码如下：

```
@RequestMapping("redirect")
public String redirectTest(Model model,SessionStatus sessionStatus){
        AyUser ayUser = new AyUser();
        ayUser.setName("ay");
        model.addAttribute("ayUser",ayUser);
        //删除 HttpSession 中的属性
        sessionStatus.setComplete();
        return "redirect:hello";
}
```

还可以设置多个对象到 HttpSession 中，具体代码如下：

```
@SessionAttributes(types = {AyUser.class, AyRole.class},value = {"ayUser",
"ayRole"})
```

type 属性用来指定放入 HttpSession 中的对象类型。

5.2.7　@ResponseBody 和@RequestBody 注解

1. @ResponseBody注解

@ResponseBody 注解用于将 Controller 方法返回的对象，通过 HttpMessageConverter 转换为指定格式后，写入到 Response 对象的 body 数据区。@Responsebody 注解将方法的返回

结果直接写入 HTTP 响应正文（ResponseBody）中，一般在异步获取数据时使用。在使用@RequestMapping 注解时，返回值通常被解析为跳转路径，在加上@Responsebody 后返回结果就不会被解析为跳转路径，而是直接写入到 HTTP 响应正文中。

使用@ResponseBody 和@RequestBody 注解之前，需要在 pom.xml 文件中引入 Jackson 相关的依赖包，具体代码如下：

```
<dependency>
    <groupId>com.fasterxml.jackson.core</groupId>
    <artifactId>jackson-core</artifactId>
    <version>2.9.5</version>
</dependency>
<dependency>
    <groupId>com.fasterxml.jackson.core</groupId>
    <artifactId>jackson-annotations</artifactId>
    <version>2.9.5</version>
</dependency>
<dependency>
    <groupId>com.fasterxml.jackson.core</groupId>
    <artifactId>jackson-databind</artifactId>
    <version>2.9.5</version>
</dependency>
```

Spring MVC 主要是利用类型转换器 messageConverters 将前台信息转换为开发者需要的格式，然后在相应的 Controller 方法接受参数前添加@RequestBody 注解，进行数据转换，或者在方法的返回值类型处添加@ResponseBody 注解，将返回信息转换成相关格式的数据。

下面看具体示例。

（1）返回普通的字符串

```
@Controller
@RequestMapping("/user")
public class AyUserController {

    @RequestMapping("/hello")
    @ResponseBody
    public String hello(){
        return "I am not view";
    }
}
```

在浏览器中输入访问路径：http://localhost:8080/user/hello，方法返回的不是视图，而是把字符串“I am not view”直接写入 HTTP 响应正文中，返回给浏览器。

（2）返回集合对象

```
@Controller
@RequestMapping("/user")
```

```
public class AyUserController {

    @RequestMapping("/hello")
    @ResponseBody
    public List<String> hello(){
        List<String> list = new ArrayList<String>();
        list.add("ay");
        list.add("al");
        return list;
    }
}
```

在浏览器输入访问路径：http://localhost:8080/user/hello，方法返回的不是视图，而是把JSON字符串“{"name":"ay","age":"3"}” 直接写入HTTP 响应正文中，返回给浏览器。

2. @RequestBody注解

@RequestBody 注解用于读取 Request 请求的 body 部分数据，使用系统默认配置的HttpMessageConverter进行解析，然后把相应的数据绑定到Controller方法的参数上。

下面看具体示例：

```
@Controller
@RequestMapping("/user")
public class AyUserController {

    @RequestMapping("/hello")
    @ResponseBody
    public void hello(@RequestBody  AyUser ayUser){
        System.out.println("name" + ayUser.getName());
        System.out.println("password" + ayUser.getPassword());
    }
}
```

5.3 信息转换详解

5.3.1 HttpMessageConverter<T>

在 Spring MVC 中，HttpMessageConverter 接口扮演着重要的角色。Spring MVC 可以接收不同的消息格式，也可以将不同的消息格式响应回去（最常见的是 JSON）。这些消息所蕴含的“有效信息”是一致的，那么各种不同的消息转换器，都会生成同样的转换结果。至于各种消息间解析细节的不同，就被屏蔽在不同的 HttpMessageConverter 实现类中了。HttpMessageConverter接口类具体的源码如下所示：

```
public interface HttpMessageConverter<T> {
```

```
    boolean canRead(Class<?> clazz, @Nullable MediaType mediaType);

          boolean canWrite(Class<?> clazz, @Nullable MediaType mediaType);

    List<MediaType> getSupportedMediaTypes();

          T read(Class<? extends T> clazz, HttpInputMessage inputMessage)
                throws IOException, HttpMessageNotReadableException;


          void write(T t, @Nullable MediaType contentType, HttpOutputMessage
outputMessage)
                throws IOException, HttpMessageNotWritableException;

    }
```

- canRead(Class<?> clazz, @Nullable MediaType mediaType)：指定转换器可以读取的对象类型，同时指定支持MIME类型（text/xml、application/json）。
- canWrite(Class<?> clazz, @Nullable MediaType mediaType)：指定转换器可以将clazz类型的对象写到响应流中，响应流支持类型在mediaType中定义。
- getSupportedMediaTypes()：该转换器支持的媒体类型。
- T read(Class<? Extends T> clazz, HttpInputMessage inputMessage)：将请求信息流转换成T类型的对象。
- write(T t, @Nullable MediaType contentType, HttpOutputMessage outputMessage)：将T类型的对象写到响应流中，同时指定响应媒体类型为contentType。

Spring MVC 为 HttpMessageConverter 接口提供了很多实现类，下面简单列举几个实现类的内容和作用，具体如表 5-8 所示。

表 5-8　HttpMessageConverter 接口实现类

属性名称	描　述
StringHttpMessageConverter	将请求信息转换为字符串，泛型 T 为 String，可读取所有媒体类型（*/*），可通过 supportedMediaTypes 属性指定媒体类型。响应信息的媒体类型为 text/plain（即 Content-Type 的值）
ByteArrayHttpMessageConverter	读写二进制数据，泛型 T 为 byte[]类型，可读取*/*，可通过 supportedMediaTypes 属性指定媒体类型，响应信息媒体类型为 application/octer-stream
MarshallingHttpMessageConverter	泛型 T 为 Object，可读取 text/xml 和 application/xml 媒体类型请求，响应信息的媒体类型为 text/xml 或 application/xml

（续表）

属性名称	描　述
Jaxb2RootElementHttpMessageConverter	通过 JAXB2 读写 XML 信息，将请求消息转换到标注 XmlRootElement 和 XmlType 注解的类中，泛型 T 为 Object，可读取 text/xml 和 application/xml 媒体类型请求，响应信息的媒体类型为 text/xml 或 application/xml
MappingJacksonHttpMessageConverter	利用 jackson 的 ObjectMapper 读写 JSON 数据，泛型 T 为 Object，可读取 application/json，响应媒体类型为 application/json
FormHttpMessageConverter	表单与 MultiValueMap 的相互转换。读支持响应 MediaType 为 application/x-www-form-urlencoded，写支持的响应类型为 application/x-www-form-urlencoded 和 multipart/form-data
SourceHttpMessageConverter	数据与 javax.xml.transform.Source 的相互转换。读支持响应 MediaType 为 text/xml 和 application/xml，写支持的响应类型为 text/xml 和 application/xml
BufferedImageHttpMessageConverter	数据与 java.awt.image.BufferedImage 的相互转换。读支持 MediaType 为 Java I/O API 的所有类型，写支持的响应类型为 Java I/O API 的所有类型

除了表 5-8 中列出的 HttpMessageConverter 接口实现类外，还有很多实现类无法一一阐述，读者可以自己阅读 Spring MVC 源码学习。

5.3.2 RequestMappingHandlerAdapter

在 Spring MVC 中，配置<mvc:annotation-driven/>标签需注册三个 bean：

- RequestMappingHandlerMapping
- RequestMappingHandlerAdapter
- DefaultHandlerExceptionResolver

DispatcherServlet 默认已经装配 RequestMappingHandlerAdapter 作为 HandlerAdapter 组件的实现类，即 HttpMessageConverter 由 RequestMappingHandlerAdapter 使用，将请求信息转换为对象，或者将对象转换为响应信息。RequestMappingHandlerAdapter 的具体源码如下：

```
    public class RequestMappingHandlerAdapter extends
AbstractHandlerMethodAdapter
        implements BeanFactoryAware, InitializingBean {
        //省略大量代码
        private List<HttpMessageConverter<?>> messageConverters;

         public RequestMappingHandlerAdapter() {
           StringHttpMessageConverter stringHttpMessageConverter
```

```
    = new StringHttpMessageConverter();
            stringHttpMessageConverter.setWriteAcceptCharset(false);  // see
SPR-7316
            this.messageConverters = new ArrayList<>(4);
            this.messageConverters.add(new ByteArrayHttpMessageConverter());
            this.messageConverters.add(stringHttpMessageConverter);
            this.messageConverters.add(new SourceHttpMessageConverter<>());
            this.messageConverters.add(new
AllEncompassingFormHttpMessageConverter());
        }
    }
```

由 RequestMappingHandlerAdapter 源码可知，RequestMappingHandlerAdapter 构造方法默认已经装配了以下的 HttpMessageConverter：

- ByteArrayHttpMessageConverter
- StringHttpMessageConverter
- SourceHttpMessageConverter
- AllEncompassingFormHttpMessageConverter

5.3.3 自定义 HttpMessageConverter

如果需要装配其他类型的 HttpMessageConverter，则可以在 Spring 的 Web 容器的上下文中自定义一个 RequestMappingHandlerAdapter。需要注意的是，如果在 Spring 的 Web 容器的上下文中自定义一个 RequestMappingHandlerAdapter，那么 Spring MVC 的 RequestMappingHandlerAdapter 默认装配的 HttpMessageConverter 将不再起作用。spring-mvc.xml 配置信息具体代码如下：

```
    <!-- 自定义 RequestMappingHandlerAdapter -->
    <bean class="org.springframework.web.servlet.mvc.method
    .annotation.RequestMappingHandlerAdapter">
       <property name="messageConverters">
       <list>
         <bean class="org.springframework.http.converter.ByteArrayHttpMessageConverter"/>
         <bean class="org.springframework.http.converter.StringHttpMessageConverter"/>
         <bean class="org.springframework.http.converter.xml.
SourceHttpMessageConverter"/>
       </list>
       </property>
    </bean>
```

<list>标签中除了使用 Spring MVC 为我们提供的 HttpMessageConverter 接口实现类外，还可以通过继承 AbstractHttpMessageConverter 实现适合自己的 HttpMessageConverter。

5.4　思考与练习

1. 请列举几个常用的请求映射注解。
2. 简述@RequestMapping 注解的作用。
3. 列举几个@RequestMapping 注解的常用属性。
4. 简单描述@RequestMapping 的 method 属性有什么作用？
5. 请列举几个常用的参数绑定注解。
6. 简述@PathVariable 注解的作用。
7. 简述@RequestParam 注解的作用。
8. 简述@CookieValue 注解的作用。
9. 简述@ResponseBody 和@RequestBody 注解的作用。
10. 动手自定义 HttMessageConverter。

第 6 章 分页开发、数据校验与事务管理

本章首先介绍 MyBatis 提供的 RowBounds 分页的使用和原理，以及分页插件 PageHelper 的使用和原理；然后介绍 Spring 的数据校验以及 Spring 和 MyBatis 事务管理。

6.1 RowBounds 类

本节将介绍 MyBatis 提供的 RowBounds 分页的使用和原理，以及分页插件 PageHelper 的使用和原理。

6.1.1 分页概述

分页是许多网站常用的功能，也是最基本的功能。分页就是将数据库查询的结果在有限的界面上分好多页显示，所以，也称为分页查询。分页查询可分为逻辑分页和物理分页。

- **逻辑分页**：依赖程序员编写的代码，数据库返回的不是分页结果，而是全部数据，然后再由程序员通过代码获取分页数据。常用的操作是一次性从数据库中查询出全部数据并存储到List集合中，因为List集合有序，再根据索引获取指定范围的数据。
- **物理分页**：使用数据库自身所带的分页机制，例如，Oracle数据库的rownum，或者MySQL数据库中的limit等机制来完成分页操作。因为是对数据库实实在在的数据进行分页条件查询，所以叫物理分页。每一次物理分页都会去连接数据库。

物理分页优于逻辑分页，因为我们没有必要将属于数据库端的压力施加到应用端来，所

以建议大家在日常工作中尽量使用物理分页而不是逻辑分页。

6.1.2　RowBounds 类

MyBatis 提供了可以进行逻辑分页的 RowBounds 类，通过传递 RowBounds 对象，来进行数据库数据的分页操作，任何 select 语句都可以使用它。然而遗憾的是该分页操作是对 ResultSet 结果集进行分页，也就是人们常说的逻辑分页而非物理分页，它会用一条 SQL 语句查询所有的结果，然后根据从第几条到第几条取出数据返回。首先，我们先来看一下 RowBounds 类的源码：

```
/**
 * @author Clinton Begin
 */
public class RowBounds {

  public static final int NO_ROW_OFFSET = 0;
  public static final int NO_ROW_LIMIT = Integer.MAX_VALUE;
  public static final RowBounds DEFAULT = new RowBounds();

  private final int offset;
  private final int limit;

  public RowBounds() {
    this.offset = NO_ROW_OFFSET;
    this.limit = NO_ROW_LIMIT;
  }

  //构造函数
public RowBounds(int offset, int limit) {
    this.offset = offset;
    this.limit = limit;
  }

  public int getOffset() {
    return offset;
  }

  public int getLimit() {
    return limit;
  }

}
```

在 RowBounds 源码中，定义了 offset 和 limit 两个参数。offset 表示从第几行开始读取数据。limit 表示限制返回的记录数。从源码还可以看出，offset 默认值为 0（NO_ROW_OFFSET = 0），limit 默认值为 Java 允许的最大整数（NO_ROW_LIMIT =

Integer.MAX_VALUE）。在数据量大的情况下，将数据一次性查询出来，这对内存消耗影响很大，会造成内存溢出的问题。

6.1.3 RowBounds 分页应用

首先，在 AyUserMapper.xml 配置文件中添加 select 查询，具体代码如下：

```
<sql id="userField">
    a.id as "id",
    a.name as "name",
    a.password as "password"
</sql>
//查询所有的用户
<select id="findAll" resultMap="userMap">
    select
    <include refid="userField"/>
    from ay_user a
</select>
```

然后，在 AyUserDao 接口中添加对应的查询方法 findAll，findAll 方法入参是 RowBounds，具体代码如下：

```
@Repository
public interface AyUserDao {
List<AyUser> findAll(RowBounds rowBounds);
    //省略其他代码
}
```

最后，在 AyUserDaoTest 测试类中添加测试方法 testFindAll()，具体代码如下：

```
/**
 * 描述：用户 DAO 测试类
 * @author Ay
 * @create 2018/05/04
 **/
public class AyUserDaoTest extends BaseJunit4Test{

    @Resource
    private AyUserDao ayUserDao;

    @Test
    public void testFindAll(){
        List<AyUser> userList = ayUserDao.findAll(new RowBounds(0, 5));
        for(AyUser ayUser: userList){
            System.out.println("name: " + ayUser.getName());
        }
    }
```

```
    }
```

代码全部开发完成之后，运行测试方法 testFindAll，便可以在控制台中看到打印信息：

```
id: 1
name: 阿兰
id: 2
name: 阿毅
id: 3
name: Ay
```

RowBounds 分页在任何 select 语句中都可以使用，但是它是在 SQL 查询出所有结果的基础上截取数据的，对于大数据量返回的情况并不适用。RowBounds 分页适合返回数据结果较少的查询。

6.1.4　RowBounds 分页原理

上一节我们已经知道 RowBounds 分页是一个逻辑分页，这一节主要学习 RowBounds 是如何来进行分页的。首先看 DefaultSqlSession 类中的查询接口，部分源码如下所示：

```
    <E> List<E> selectList(String statement, Object parameter);

    <E> List<E> selectList(String statement, Object parameter, RowBounds
rowBounds);
```

从源码可以看出，DefaultSqlSession 提供的查询接口是以 RowBounds 作为参数用来进行分页的。

再来看 DefaultResultSetHandler 类的源码，该类主要是对查询结果集进行处理的类，部分源码如下：

```
    private void handleRowValuesForSimpleResultMap(ResultSetWrapper rsw,
            ResultMap resultMap,ResultHandler<?> resultHandler,
            RowBounds rowBounds,ResultMapping parentMapping)throws SQLException
{

        DefaultResultContext<Object> resultContext
                                        = new DefaultResultContext<Object>();
        //跳到 offset 位置，准备读取数据
    skipRows(rsw.getResultSet(), rowBounds);
        //while 循环判断是否小于 limit 值，如果是，读取 limit 条数据
    while (shouldProcessMoreRows(resultContext, rowBounds)
                                            && rsw.getResultSet().next()) {
        ResultMap discriminatedResultMap =
                resolveDiscriminatedResultMap(rsw.getResultSet(), resultMap,
null);
        Object rowValue = getRowValue(rsw, discriminatedResultMap);
        storeObject(resultHandler, resultContext,
```

```
                        rowValue, parentMapping, rsw.getResultSet());
        }
    }

    private void skipRows(ResultSet rs, RowBounds rowBounds) throws
SQLException {
        if (rs.getType() != ResultSet.TYPE_FORWARD_ONLY) {
          if (rowBounds.getOffset() != RowBounds.NO_ROW_OFFSET) {
            //直接定位
    rs.absolute(rowBounds.getOffset());
          }
        } else {
          //通过循环，跳到offset位置，进行读取
          for (int i = 0; i < rowBounds.getOffset(); i++) {
            rs.next();
          }
        }
    }
```

从源码可以看出，RowBounds 在处理分页时只是简单地把 offset 之前的数据都 skip 掉，超过 limit 之后的数据不取出。跳过 offset 之前的数据是由方法 skipRows 处理，判断数据是否超过了 limit 则是由 shouldProcessMoreRows 方法进行判断。简单来说，就是先把数据全部查询到 ResultSet，然后从 ResultSet 中取出 offset 和 limit 之间的数据，实现分页查询。

6.1.5 分页插件 PageHelper

PageHelper 是一款开源免费的 MyBatis 物理分页插件。PageHelper 插件可以方便地实现物理分页，与 RowBounds 分页方式相比，PageHelper 在查询性能方面，更胜一筹。PageHelper 的 github 地址：https://github.com/pagehelper/Mybatis-PageHelper，读者可访问该地址下载相关的文档和资料。

PageHelper 使用非常简单，首先，在项目的 pom.xml 文件添加 PageHelper 依赖包，具体代码如下：

```
<dependency>
     <groupId>com.github.pagehelper</groupId>
     <artifactId>pagehelper</artifactId>
     <version>5.1.4</version>
</dependency>
```

依赖添加完成之后，在 applicationContext.xml 配置文件中添加 PageHelper 相关配置，具体代码如下：

```
//省略代码
!--3. 配置SqlSessionFactory对象-->
<bean id="sqlSessionFactory"
```

```
class="org.mybatis.spring.SqlSessionFactoryBean">
        <!--注入数据库连接池-->
        <property name="dataSource" ref="dataSource"/>
        <!--扫描 sql 配置文件:mapper 需要的 xml 文件-->
        <property name="mapperLocations" value="classpath:mapper/*.xml"/>

        <!-- 配置分页插件 -->
        <property name="plugins">
            <array>
                <bean class="com.github.pagehelper.PageInterceptor">
                    <property name="properties">
                        <value>
                            <!-- 数据库类型为 mysql -->
                            helperDialect=mysql
                            <!-- 启用合理化时，如果 pageNum <1 会查询第一页，如果 pageNum
> pages 会查询最后一页 -->
    <!-- 禁用合理化时，如果 pageNum < 1 或 pageNum > pages 会返回空数据 -->
                            reasonable=true
                        </value>
                    </property>
                </bean>
            </array>
        </property>
    </bean>
    //省略代码
```

配置添加完成之后，在 AyUserDaoTest 开发测试用例，具体代码如下：

```
    @Test
    public void testPageHelper(){
        //startPage(第几页, 多少条数据)
        PageHelper.startPage(0, 1);
        //查询所有用户
        List<AyUser> userList = ayUserDao.findAll();
        //用 PageInfo 对结果进行包装
        PageInfo pageInfo = new PageInfo(userList);
    }
```

PageHelper 使用非常简单，在需要进行分页的 MyBatis 方法前调用 PageHelper.startPage 静态方法即可，紧跟在这个方法后的第一个 MyBatis 查询方法会被进行分页，然后分页插件会把分页信息封装到 PageInfo 中。PageInfo 包含了非常全面的分页属性，PageInfo 具体源码如下：

```
    public class PageInfo<T> extends PageSerializable<T> {
        //当前页
        private int pageNum;
```

```
    //每页的数量
    private int pageSize;
    //当前页的数量
    private int size;

    //由于 startRow 和 endRow 不常用，这里介绍一个具体的用法
    //可以在页面中"显示 startRow 到 endRow 共 size 条数据"

    //当前页面第一个元素在数据库中的行号
    private int startRow;
    //当前页面最后一个元素在数据库中的行号
    private int endRow;
    //总页数
    private int pages;

    //前一页
    private int prePage;
    //下一页
    private int nextPage;

    //是否为第一页
    private boolean isFirstPage = false;
    //是否为最后一页
    private boolean isLastPage = false;
    //是否有前一页
    private boolean hasPreviousPage = false;
    //是否有下一页
    private boolean hasNextPage = false;
    //导航页码数
    private int navigatePages;
    //所有导航页号
    private int[] navigatepageNums;
    //导航条上的第一页
    private int navigateFirstPage;
    //导航条上的最后一页
    private int navigateLastPage;
    //省略代码
}
```

6.2 Spring 数据校验

本节将介绍 Spring 的 Validation 校验框架、JSR 303 校验以及常用的注解。

6.2.1 数据校验概述

在 Web 应用程序中，为了防止客户端传来的数据引发程序异常，常常需要对数据进行验证。数据验证分为客户端验证与服务器端验证。客户端验证主要通过 JavaScript 脚本进行，而服务器端验证则主要通过 Java 代码进行验证。为了保证数据的安全性，客户端和服务器端验证都是必须的。

Spring MVC 提供了强大的数据校验功能，其中有两种方法可以验证输入：

（1）利用 Spirng 自带的 Validation 校验框架。

（2）利用 JSR 303（Java 检验规范）实现校验功能。

下面详细介绍两种校验方法。

6.2.2 Spring 的 Validation 校验框架

Spring 提供了 Validator 接口来校验对象，Validator 接口源码如下所示：

```
public interface Validator {
        boolean supports(Class<?> clazz);
        void validate(@Nullable Object target, Errors errors);
}
```

- boolean supports(Class<?> clazz)：校验器能够对clazz类型的对象进行校验。
- validate(@Nullable Object target, Errors errors)：对目标类target进行校验，并将校验错误记录在errors中。
- Errors类：用来存放错误信息的接口。Errors对象包含一系列的FieldError和ObjectError对象。FieldError表示与被获取的对象中的某一属性相关的一个错误。

除了 Validator 和 Errors 对象之外，Spring 的 Validation 校验框架还提供了其他重要的接口：

- ValidationUtils：校验工具类，提供多个给Errors对象保存错误的方法。
- LocalValidatorFactoryBean：该类实现了Spring的Validator接口，也实现了JSR 303的Validator接口。

在实际开发过程中，spring-mvc.xml 配置文件中的<mvc:annotation-driven/>会默认装配好 LocalValidatorFactoryBean，所以在实际开发过程中不需要手动配置 LocalValidatorFactoryBean。下面来看一个具体的实例。

首先，在 src\main\java\com\ay\model 目录下创建 AyUser 实体类（之前章节已创建），具体代码如下：

```
/**
 * 用户实体
 * @author Ay
```

```
 * @date  2018/04/02
 */
public class AyUser implements Serializable{

    private Integer id;
    private String name;
    private String password;
    private Integer age;
    //省略 set、get 方法
}
```

其次，在 src\main\java\com\ay\validator 目录下创建用户数据校验类 AyUserValidator，具体代码如下：

```
/**
 * 描述：用户数据校验类
 * @author Ay
 * @create 2018/05/25
 **/
@Component
public class AyUserValidator implements Validator {

    /**
     * This Validator validates *just* AyUser instances
     */
    public boolean supports(Class clazz) {
        return AyUser.class.equals(clazz);
    }

    public void validate(Object obj, Errors e) {
        //指定 errors 对象、验证失败的字段、错误码
        ValidationUtils.rejectIfEmpty(e, "name", "name.empty");
        AyUser p = (AyUser) obj;
        if (p.getAge() < 0) {
            e.rejectValue("age", "年龄不能小于 0 岁");
        } else if (p.getAge() > 150) {
            e.rejectValue("age", "年龄不能超过 150 岁");
        }
    }
}
```

AyUserValidator 类实现 Validator 接口，并实现 supports()和 validate()方法。在 supports()方法中使用 ValidationUtils 的静态方法 rejectIfEmpty()对 name 和 age 属性进行校验。假如 name 属性是 null 或者空字符串，就拒绝验证通过。假如 age 属性的值年龄小于 0 岁或者大于 150 岁，就拒绝验证通过。通过使用 Errors 对象的 rejectValue 方法保存校验信息。

最后，在 src\main\webapp\WEB-INF\views 目录下创建保存用户页面 saveUser.jsp，具体代码如下：

```
<%@page language="java" contentType="text/html; charset=UTF-8"
        pageEncoding="UTF-8" isELIgnored="false"%>
<!DOCTYPE HTML>
<html>
<head>
    <title>Getting Started: Serving Web Content</title>
    <meta http-equiv="Content-Type" content="text/html; charset=UTF-8" />
</head>
<body>
<form method="post"  action="/user/insert" >
    <table>
        <tr>
            <td>姓名：</td>
            <td><input id="name" name="name" type="text"></td>
        </tr>
        <tr>
            <td>密码：</td>
            <td><input id="password" name="password" type="text"></td>
        </tr>
        <tr>
            <td>年龄：</td>
            <td><input id="age" name="age" type="text"></td>
        </tr>
        <tr>
            <td><input type="submit" value="提交"></td>
        </tr>
    </table>
</form>
</body>
<script></script>
</html>
```

页面 saveUser.jsp 创建完成之后，在控制层 AyUserController 添加相关的代码，具体如下：

```
@Controller
@RequestMapping("/user")
public class AyUserController {

    @Resource
    private AyUserValidator ayUserValidator;

    @RequestMapping("/save")
```

```
    public String save(){
        return "saveUser";
    }

    @PostMapping("/insert")
    public String insert(@ModelAttribute AyUser ayUser,Model model, Errors
errors){
        ayUserValidator.validate(ayUser, errors);
        if(errors.hasErrors()){
            //将错误存放到model中
            model.addAttribute("errors", errors);
            return "error";
        }
        int count = ayUserService.insert(ayUser);
        return "success";
    }
}
```

在控制层类 AyUserController 中，注入 AyUserValidator 校验类，同时在 insert 方法中调用 AyUserValidator 类的 validate 方法进行校验。如果数据校验正确，返回成功（success）界面；如果数据校验失败，返回错误（error）界面。

success.jsp 页面如下所示：

```
<%@page language="java" contentType="text/html; charset=UTF-8"
        pageEncoding="UTF-8" isELIgnored="false"%>
<!DOCTYPE HTML>
<html>
<head>
    <title>Getting Started: Serving Web Content</title>
    <meta http-equiv="Content-Type" content="text/html; charset=UTF-8" />
</head>
<body>
success !!!
</body>
<script></script>
</html>
```

error.jsp 页面如下所示：

```
<%@page language="java" contentType="text/html; charset=UTF-8"
        pageEncoding="UTF-8" isELIgnored="false"%>
<!DOCTYPE HTML>
<html>
<head>
    <title>Getting Started: Serving Web Content</title>
    <meta http-equiv="Content-Type" content="text/html; charset=UTF-8" />
```

```
</head>
<body>
error !!!
</body>
<script></script>
</html>
```

代码开发完成之后，重新启动 Tomcat 服务器，项目启动成功后，在浏览器中输入访问 URL：http://localhost:8080/user/save，出现保存用户界面，具体如图 6-1 所示。

在保存用户页面中填写用户信息（姓名：ay；密码：123；年龄：199），然后单击“保存”按钮，由于年龄不符合数据校验规则，故返回到错误界面，具体如图 6-2 所示。

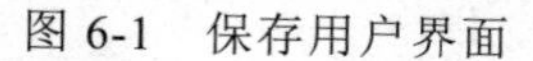

图 6-1　保存用户界面

图 6-2　保存错误界面

由于 Spring Validation 框架通过硬编码完成数据校验，在实际开发过程中会显得比较麻烦，因此更加推荐使用 JSR 303 完成数据校验。

6.2.3　JSR 303 校验

JSR 303 是 Java 为 Bean 数据合法性校验提供的标准框架，已经包含在 Java EE 6.0 中。JSR 是一个规范，它的核心接口是 Validator，该接口根据目标对象类中所标注的校验注解进行数据校验，并得到校验结果。JSR 303 通过在 Bean 属性中标注类似@NotNull、@Max 等标准的注解指定校验规则，并通过标准的验证接口对 Bean 进行验证。JSR 303 包含的详细注解如表 6-1 所示。

表 6-1　JSR 303 包含注解

注解名称	描　　述
@Null	被注释的元素必须为 null
@NotNull	被注释的元素必须不为 null
@AssertTrue	被注释的元素必须为 true
@AssertFalse	被注释的元素必须为 false
@Min(value)	被注释的元素必须是一个数字，其值必须大于等于指定的最小值
@Max(value)	被注释的元素必须是一个数字，其值必须小于等于指定的最大值
@DecimalMin(value)	被注释的元素必须是一个数字，其值必须大于等于指定的最小值
@DecimalMax(value)	被注释的元素必须是一个数字，其值必须小于等于指定的最大值

（续表）

注解名称	描　述
@Size(max=, min=)	被注释的元素的大小必须在指定的范围内
@Digits (integer, fraction)	被注释的元素必须是一个数字，其值必须在可接受的范围内
@Past	被注释的元素必须是一个过去的日期
@Future	被注释的元素必须是一个将来的日期
@Pattern(regex=,flag=)	被注释的元素必须符合指定的正则表达式

HIbernate Validator 是 JSR 303 的一个参考实现，除了支持所有标准的校验注解外，还支持一些扩展注解，具体内容如表 6-2 所示。

表 6-2　Hibernate Validator 附加的注解

注解名称	描　述
@NotBlank(message =)	验证字符串非 null，且长度必须大于 0
@Email	被注释的元素必须是电子邮箱地址
@Length(min=,max=)	被注释的字符串的大小必须在指定的范围内
@NotEmpty	被注释的字符串必须非空
@Range(min=,max=,message=)	被注释的元素必须在合适的范围内
@URL	被注释的元素必须是合法的 URL

下面来看一个具体的示例。

首先，在 pom.xml 文件中引入 Hibernate Validator 所需要的依赖包，具体代码如下：

```
<dependency>
     <groupId>org.hibernate.validator</groupId>
     <artifactId>hibernate-validator</artifactId>
     <version>6.0.10.Final</version>
</dependency>
```

其次，在 src\main\webapp\WEB-INF\views 目录下创建保存用户页面 saveUser.jsp 和 AyUser.java。

AyUser.java 实体类的具体代码如下：

```
/**
 * 用户实体
 * @author Ay
 * @date  2018/04/02
 */
public class AyUser implements Serializable{

   private Integer id;
   @NotBlank(message = "name 不能为空")
   private String name;
```

```
    @Length(min = 3, max = 16, message = "密码长度必须在 3~16 位之间")
    private String password;
    @Range(min = 0, max = 150, message = "年龄必须在 0~150 岁之间")
    private Integer age;
    //省略 set、get 方法
}
```

saveUser.jsp 的具体代码如下：

```
<%@page language="java" contentType="text/html; charset=UTF-8"
      pageEncoding="UTF-8" isELIgnored="false"%>
<!DOCTYPE HTML>
<html>
<head>
   <title>Getting Started: Serving Web Content</title>
   <meta http-equiv="Content-Type" content="text/html; charset=UTF-8" />
</head>
<body>
<form method="post"  action="/user/insert" >
   <table>
      <tr>
         <td>姓名：</td>
         <td><input id="name" name="name" type="text"></td>
      </tr>
      <tr>
         <td>密码：</td>
         <td><input id="password" name="password" type="text"></td>
      </tr>
      <tr>
         <td>年龄：</td>
         <td><input id="age" name="age" type="text"></td>
      </tr>
      <tr>
         <td><input type="submit" value="保存"></td>
      </tr>
   </table>
</form>
</body>
<script></script>
</html>
```

最后，在 AyUserController 类中添加相关的方法，具体代码如下：

```
/**
 * 用户控制层
 *@author Ay
 * @date 2018/04/02
```

```
 */
@Controller
@RequestMapping("/user")
public class AyUserController {

    @RequestMapping("/save")
    public String save(){
        return "saveUser";
    }

    @PostMapping("/insert")
    public String insert(@Valid AyUser ayUser, Model model, BindingResult
result){
        if(errors.hasErrors()){
            model.addAttribute("errors", errors);
            return "error";
        }
        int count = ayUserService.insert(ayUser);
        return "success";
    }
}
```

- BindingResult：BindingResult扩展了Errors接口，同时可以获取数据绑定结果对象的信息。@Valid和BindingResult参数是成对出现的，并且在形参中出现的顺序是固定的，一前一后。

在 AyUserController 类中 insert()方法使用@Valid 注解对提交的数据进行校验，同时 insert()方法使用 Errors 对象保存校验信息。代码开发完成之后，在浏览器输入访问路径：http://localhost:8080/user/save，可以看到保存用户界面，具体如图 6-3 所示。

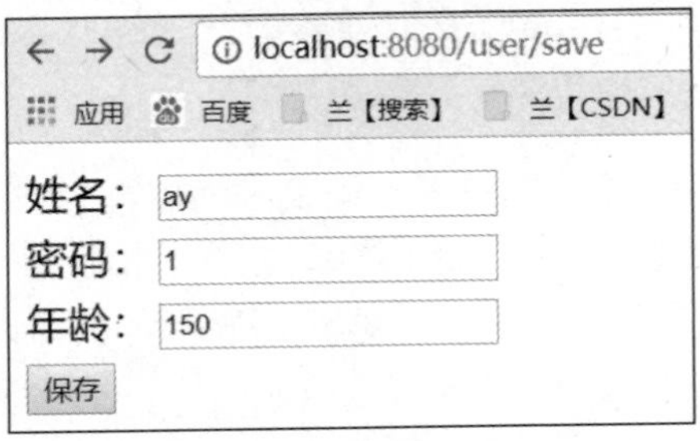

图 6-3　保存用户界面

在保存用户界面中，填写相关的用户信息（姓名：ay；密码：1；年龄：150）并单击“保存”按钮，由于密码的长度要求是 3~16 位，故可以在控制台看到数据校验错误信息，具体如图 6-4 所示。

```
Field error in object 'ayUser' on field 'password': rejected value [];
codes [Length.ayUser.password,Length.password,Length.java.lang.String,Length];
arguments [org.springframework.context.support.DefaultMessageSourceResolvable:
codes [ayUser.password,password]; arguments []; default message [password],16,3];
default message [密码长度必须在3 16位之间]
        at org.springframework.web.method.annotation.ModelAttributeMethodProcessor
        at org.springframework.web.method.support.HandlerMethodArgumentResolverCom
        at org.springframework.web.method.support.InvocableHandlerMethod.getMethod
        at org.springframework.web.method.support.InvocableHandlerMethod.invokeFor
        at org.springframework.web.servlet.mvc.method.annotation.ServletInvocableH
        at org.springframework.web.servlet.mvc.method.annotation.RequestMappingHan
        at org.springframework.web.servlet.mvc.method.annotation.RequestMappingHan
        at org.springframework.web.servlet.mvc.method.AbstractHandlerMethodAdapter
        at org.springframework.web.servlet.DispatcherServlet.doDispatch(Dispatcher
        at org.springframework.web.servlet.DispatcherServlet.doService(DispatcherS
        at org.springframework.web.servlet.FrameworkServlet.processRequest(Framewo
```

图 6-4　用户校验错误信息

6.3 Spring 和 MyBatis 事务管理

本节将介绍 Spring 事务管理（包括 Spring 声明式事务、Spring 注解事务行为）和 MyBatis 事务管理。

6.3.1 Spring 事务管理

1. Spring事务回顾

事务管理是企业级应用程序开发中必不可少的技术，用来确保数据的完整性和一致性。事务有 4 大特性（ACID）：原子性（atomicity）、一致性（consistency）、隔离性（isolation）和持久性（durability）。作为企业级应用程序框架，Spring 在不同的事务管理 API 之上定义了一个抽象层，而应用程序开发人员不必了解底层的事务管理 API，就可以使用 Spring 的事务管理机制。

Spring 既支持编程式事务管理（也称编码式事务），也支持声明式的事务管理。编程式事务管理是指将事务管理代码嵌入到业务方法中来控制事务的提交和回滚。在编程式事务中，必须在每个业务操作中包含额外的事务管理代码。声明式事务管理是指将事务管理代码从业务方法中分离出来，以声明的方式来实现事务管理。大多数情况下声明式事务管理比编程式事务管理更好用。Spring 通过 Spring AOP 框架支持声明式事务管理。

数据访问的技术很多，如 JDBC、JPA、Hibernate、分布式事务等，面对众多的数据访问技术，Spring 并不直接管理事务，而是提供了许多内置事务管理器实现，常用的有：DataSourceTransactionManager、JdoTransactionManager、JpaTransactionManager 及 HibernateTransactionManager 等。

2. Spring声明式事务

Spring 配置文件中关于事务配置总是由三个组成部分，分别是 DataSource、TransactionManager 和代理机制。无论哪种配置方式，一般变化的只是代理机制这部分，DataSource 和 TransactionManager 这两部分只会根据数据访问方式有所变化，比如使用 Hibernate 进行数据访问时，DataSource 实现为 SessionFactory，TransactionManager 的实现为 HibernateTransactionManager。

Spring 声明式事务配置提供 5 种方式，而基于 Annotation 注解方式目前比较流行，所以这里只简单介绍基于注解方式配置 Spring 声明式事务。可以使用@Transactional 注解在类或者方法上表明该类或者方法需要事务支持，被注解的类或者方法被调用时，Spring 开启一个新的事务，当方法正常运行时，Spring 会提交这个事务。具体例子如下：

```
@Transactional
    public AyUser update() {
        //执行数据库操作
}
```

同时需要在 applicationContext.xml 文件中添加事务相关的配置，具体代码如下：

```
<!-- 声明式事务 -->
<tx:annotation-driven transaction-manager="transactionManager"
                    proxy-target-class="true" />
<bean id="transactionManager"
        class="org.springframework.jdbc.datasource.
DataSourceTransactionManager">
    <property name="dataSource" ref="dataSource" />
</bean>
```

这里需要注意的是，@Transactional 注解来自 org.springframework.transaction.annotation，Spring 提供了@EnableTransactionManagement 注解在配置类上来开启声明式事务的支持，使用@EnableTransactionManagement 后，Spring 容器会自动扫描注解@Transactional 的方法和类。

3. Spring注解事务行为

当事务方法被另一个事务方法调用时，必须指定事务应该如何传播。例如，方法可能继续在现有事务中运行，也可能开启一个新的事务，并在自己的事务中运行。事务的传播行为可以在@Transactional 的属性中指定，Spring 定义了 7 种传播行为，具体如表 6-3 所示。

表 6-3　Spring 传播行为

传播行为	含　义
PROPAGATION_REQUIRED	如果当前没有事务，就新建一个事务，如果已经存在一个事务，加入到这个事务中
PROPAGATION_SUPPORTS	支持当前事务，如果当前没有事务，就以非事务方式执行
PROPAGATION_MANDATORY	使用当前的事务，如果当前没有事务，就抛出异常
PROPAGATION_REQUIRES_NEW	新建事务，如果当前存在事务，把当前事务挂起
PROPAGATION_NOT_SUPPORTED	以非事务方式执行操作，如果当前存在事务，就把当前事务挂起
PROPAGATION_NEVER	以非事务方式执行，如果当前存在事务，则抛出异常
PROPAGATION_NESTED	如果当前存在事务，则在嵌套事务内执行。如果当前没有事务，则执行与 PROPAGATION_REQUIRED 类似的操作

隔离级别定义了一个事务可能受其他并发事务影响的程度。在典型的应用程序中，多个事务并发运行，经常会操作相同的数据来完成各自的任务。并发虽然是必须的，但也可能会导致许多问题，并发事务所导致的问题可以分为以下三类：

- 脏读（Dirty reads）：脏读发生在一个事务读取了另一个事务改写但尚未提交的数据。如果改写在稍后被回滚了，那么第一个事务获取的数据就是无效的。
- 不可重复读（Nonrepeatable read）：不可重复读发生在一个事务执行相同的查询两次或两次以上，但是每次都得到不同的数据时。这通常是因为另一个并发事务在两次查询期间更新了数据。
- 幻读（Phantom read）：幻读与不可重复读类似。它发生在一个事务（T1）读取了几行数据，接着另一个并发事务（T1）插入了一些数据时。在随后的查询中，第一个事务（T1）就会发现多了一些原本不存在的记录。

针对这些问题，Spring提供了5种事务的隔离级别，具体如表6-4所示。

表6-4 Spring隔离级别

隔离级别	含 义
ISOLATION_DEFAULT	使用数据库默认的事务隔离级别，另外4个与JDBC的隔离级别相对应
ISOLATION_READ_UNCOMMITTED	事务最低的隔离级别，允许读取尚未提交的更改。可能导致脏读、幻读或不可重复读
ISOLATION_READ_COMMITTED	允许从已经提交的并发事务中读取。可防止脏读，但幻读和不可重复读仍可能会发生
ISOLATION_REPEATABLE_READ	对相同字段的多次读取的结果是一致的，除非数据被当前事务本身改变。可防止脏读和不可重复读，但幻读仍可能发生
ISOLATION_SERIALIZABLE	完全服从ACID的隔离级别，确保不发生脏读、不可重复读和幻读。这种隔离级别是最慢的，因为它通常是通过完全锁定当前事务所涉及的数据表来完成的

@Transactional可以通过propagation属性定义事务行为，属性值分别为：REQUIRED、SUPPORTS、MANDATORY、REQUIRES_NEW、NOT_SUPPORTED、NEVER及NESTED，分别对应表5-1中的内容。可以通过isolation属性定义隔离级别，属性值分别为：DEFAULT、READ_UNCOMMITTED、READ_COMMITTED、REPEATABLE_READ及SERIALIZABLE。

还可以通过timeout属性设置事务过期时间，通过readOnly指定当前事务是否是只读事务，通过rollbackFor（noRollbackFor）指定哪个或者哪些异常可以引起（或不可以引起）事务回滚。

6.3.2 MyBatis事务管理

MyBatis使用Transaction接口对数据库事务进行了抽象，Transaction接口的定义如下：

```
public interface Transaction {
```

```
    //获取数据库连接对象
    Connection getConnection() throws SQLException;
       //提交事务
    void commit() throws SQLException;
       //回滚事务
    void rollback() throws SQLException;
       //关闭数据库连接
    void close() throws SQLException;
       //获取事务超时时间
    Integer getTimeout() throws SQLException;

  }
```

Transaction 接口有两个实现类，即 JdbcTransaction 和 ManagedTransaction，分别由 JdbcTransactionFactory 和 ManagedTransactionFactory 负责创建，具体如图 6-5 所示。

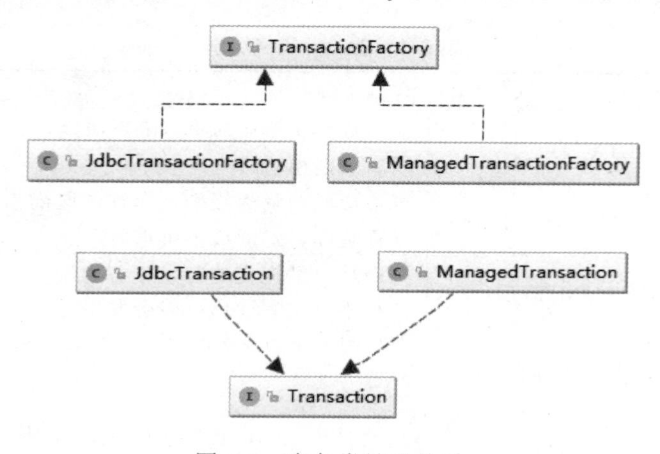

图 6-5　事务类继承关系

JdbcTransaction 依赖 JDBC 的 Connection 控制事务的提交和回滚，JdbcTransaction 源码如下所示：

```
   public class JdbcTransaction implements Transaction {
     //省略代码
     // connection 对象会延迟初始化
   protected Connection connection;

     protected DataSource dataSource;
     protected TransactionIsolationLevel level;
     // MEMO: We are aware of the typo. See #941
     protected boolean autoCommmit;

     public JdbcTransaction(DataSource ds, TransactionIsolationLevel
desiredLevel,
   boolean desiredAutoCommit) {
```

```
    dataSource = ds;
    level = desiredLevel;
    autoCommmit = desiredAutoCommit;
  }

  @Override
  public void commit() throws SQLException {
    if (connection != null && !connection.getAutoCommit()) {
      if (log.isDebugEnabled()) {
        log.debug("Committing JDBC Connection [" + connection + "]");
      }
      //使用 connection 的 commit 功能
      connection.commit();
    }
  }

  @Override
  public void rollback() throws SQLException {
    if (connection != null && !connection.getAutoCommit()) {
      if (log.isDebugEnabled()) {
        log.debug("Rolling back JDBC Connection [" + connection + "]");
      }
      //使用 connection 的 rollback 功能
      connection.rollback();
    }
  }

  @Override
  public void close() throws SQLException {
    if (connection != null) {
      resetAutoCommit();
      if (log.isDebugEnabled()) {
        log.debug("Closing JDBC Connection [" + connection + "]");
      }
      //使用 connection 的 rollback 功能
      connection.close();
    }
  }
}
```

由上述代码可知，JdbcTransaction 直接使用 JDBC 的提交和回滚事务机制，它依赖于从 dataSource 中取得 connection 连接来管理事务。JdbcTransaction 只是对 java.sql.Connection 事务进行了再次封装，JdbcTransaction 对象在初始化时，构造方法会初始化 DataSource、TransactionIsolationLevel 和 boolean desiredAutoCommit 字段，而 connection 字段会延迟初始化，它会在调用 getConnection()方法时，通过 dataSurce.getConnection()方法初始化。

ManagedTransaction 同样是依赖 dataSource 字段获取连接，但是其 commit()、rollback() 方法都是空实现，事务的提交和回滚都是依靠容器管理的。ManagedTransaction 通过 closeCollection()字段的值控制数据库连接的关闭行为。ManagedTransaction 的源码如下所示：

```
public class ManagedTransaction implements Transaction {

  private static final Log log = LogFactory.getLog(ManagedTransaction.class);
  private DataSource dataSource;
  private TransactionIsolationLevel level;
  private Connection connection;
  private final boolean closeConnection;
  //构造方法
  public ManagedTransaction(Connection connection, boolean
closeConnection) {
    this.connection = connection;
    this.closeConnection = closeConnection;
  }

  public ManagedTransaction(DataSource ds, TransactionIsolationLevel level,
boolean closeConnection) {
    this.dataSource = ds;
    this.level = level;
    this.closeConnection = closeConnection;
  }

  @Override
  public Connection getConnection() throws SQLException {
    if (this.connection == null) {
      openConnection();
    }
    return this.connection;
  }
  //空实现
  @Override
  public void commit() throws SQLException {
    // Does nothing
  }
  //空实现
  @Override
  public void rollback() throws SQLException {
    // Does nothing
  }

  @Override
```

```
    public void close() throws SQLException {
      if (this.closeConnection && this.connection != null) {
        if (log.isDebugEnabled()) {
          log.debug("Closing JDBC Connection [" + this.connection + "]");
        }
        this.connection.close();
      }
    }

    protected void openConnection() throws SQLException {
      if (log.isDebugEnabled()) {
        log.debug("Opening JDBC Connection");
      }
      this.connection = this.dataSource.getConnection();
      if (this.level != null) {
        this.connection.setTransactionIsolation(this.level.getLevel());
      }
    }

    @Override
    public Integer getTimeout() throws SQLException {
      return null;
    }

  }
```

6.4　思考与练习

1. 简述逻辑分页与物理分页的区别。
2. 物理分页为什么优于逻辑分页？
3. 简述 RowBounds 分页的原理。
4. RowBounds 分页属于物理分页还是逻辑分页？
5. 使用分页插件 PageHelper 完成分页功能。
6. Spring MVC 提供哪两种数据校验功能。
7. 简单描述 JSR 303 校验？
8. 简述 JSR 303 注解@NotNull 含义。
9. 简述 JSR 303 注解@Max(value)和@Min(value)的含义。
10. JSR 303 提供了哪些附加的注解。
11. 简述事务的 4 大特性。
12. Spring 事务配置由哪几部分组成？
13. 简述@EnableTransactionManagement 注解的作用。

14. Spring 定义了哪几种事务传播行为？
15. 简述 PROPAGATION_REQUIRED 事务传播行为的含义。
16. 简述并发事务会导致哪几种问题？
17. 简单描述脏读的含义。
18. 简单描述 Spring 提供的 5 种隔离级别。
19. Transaction 接口提供了哪些方法？

第 7 章 MyBatis 缓存机制

本章将介绍 MyBatis 缓存机制，包括一级缓存和二级缓存以及一级缓存和二级缓存的使用及原理。

7.1 MyBatis 的缓存模式

缓存在互联网系统中是非常重要的，其主要作用是将数据保存到内存中，当用户查询数据时，优先从缓存容器中获取数据，而不是频繁地从数据库中查询数据，从而提高查询性能。目前流行的缓存服务器有 MongoDB、Redis、Ehcache 等，不同的缓存服务器有不同的应用场景，不存在孰优孰劣。

MyBatis 提供了一级缓存和二级缓存的机制。一级缓存是 SqlSession 级别的缓存，在操作数据库时，每个 SqlSession 类的实体对象中都有一个 HashMap 数据结构可以用来缓存数据，不同的 SqlSession 类的实例对象缓存的 HashMap 数据结构互不影响。二级缓存是 Mapper 级别的缓存，多个 SqlSession 类的实例对象操作同一个 Mapper 配置文件中的 SQL 语句，可以共用二级缓存，二级缓存是跨 SqlSession 的。这里需要注意的是，在没有配置的默认情况下，MyBatis 只开启一级缓存。

MyBatis 的缓存模式如图 7-1 所示。

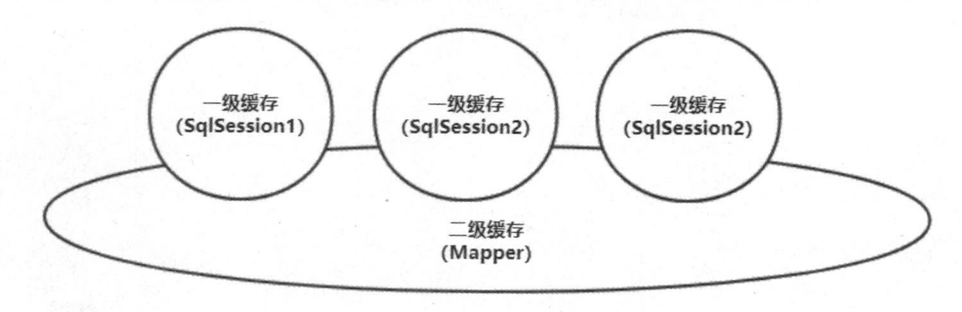

图 7-1 MyBatis 缓存模式

7.2 一级查询缓存

7.2.1 一级缓存概述

MyBatis 的一级缓存是 SqlSession 级别的缓存，在操作数据库时需要构造 SqlSession 对象。每个 SqlSession 对象中都有一个 HashMap 对象用于缓存数据，不同的 SqlSession 之间的缓存数据区域互不影响。

MyBatis 的一级缓存作用域是 SqlSession 范围的，在参数和 SQL 完全一样的情况下，使用同一个 SqlSession 对象调用同一个 Mapper 方法，往往只执行一次 SQL。因为 MyBatis 会将数据放在缓存中，下次查询的时候，如果没有声明需要刷新缓存并且缓存没有超时，SqlSession 都只会取出当前缓存的数据，而不会再次发送 SQL 到数据库中。需要注意的是，如果 SqlSession 执行了 DML 操作（insert、update 和 delete）并提交到数据库，MyBatis 会清空 SqlSession 中的一级缓存，这样做的目的是为了保证缓存中存储的是最新的信息，以避免出现脏读现象。MyBatis 的缓存机制是基于 id 进行缓存的，MyBatis 使用 HashMap 缓存数据时，使用对象的 id 作为 key，而对象则作为 value 保存。

7.2.2 一级缓存示例

下面来看 MyBatis 一级缓存的应用实例，具体代码如下：

```
    @Resource
    private SqlSessionFactoryBean sqlSessionFactoryBean;

    @Test
    public void testSessionCache() throws Exception{
            SqlSessionFactory sqlSessionFactory =
sqlSessionFactoryBean.getObject();
            SqlSession sqlSession = sqlSessionFactory.openSession();
            AyUserDao ayUserDao = sqlSession.getMapper(AyUserDao.class);
            //第一次查询
            AyUser ayUser = ayUserDao.findById("1");
```

```
        System.out.println("name: " + ayUser.getName()
                + "  password:" + ayUser.getPassword());
        //第二次查询
        AyUser ayUser2 = ayUserDao.findById("1");
        System.out.println("name: " + ayUser2.getName()
            + "  password:" + ayUser2.getPassword());
        sqlSession.close();;
    }
```

上述代码中，通过@Resource注解注入SqlSessionFactoryBean对象，SqlSessionFactoryBean对象在applicationContext.xml配置文件中已配置，具体代码如下：

```
<!--2. 数据源 druid -->
<bean id="dataSource" class="com.alibaba.druid.pool.DruidDataSource"
init-method="init" destroy-method="close">
    <property name="driverClassName" value="${jdbc.driverClassName}" />
    <property name="url" value="${jdbc.url}" />
    <property name="username" value="${jdbc.username}" />
    <property name="password" value="${jdbc.password}" />
</bean>
<!--3. 配置 SqlSessionFactory 对象-->
<bean id="sqlSessionFactory"
class="org.mybatis.spring.SqlSessionFactoryBean">
    <!--注入数据库连接池-->
    <property name="dataSource" ref="dataSource"/>
    <!--扫描 sql 配置文件:mapper 需要的 xml 文件-->
    <property name="mapperLocations" value="classpath:mapper/*.xml"/>
</bean>
```

通过SqlSessionFactory工厂获取SqlSession对象，通过SqlSession对象的getMapper()方法获取AyUserDao接口对象，并执行AyUserDao接口对象的findById()方法。AyUserDao和AyUserMapper.xml的代码如下：

```
@Repository
public interface AyUserDao {

    AyUser findById(String id);
}

<select id="findById" parameterType="String" resultMap="userMap">
    SELECT * FROM ay_user
    WHERE id = #{id}
</select>
```

执行测试用例testSessionCache()，控制台打印相关的信息，具体如图7-2所示。

```
11:47:07,050 DEBUG DataSourceUtils:114 - Fetching JDBC Connection from DataSource
Thu May 24 11:47:07 CST 2018 WARN: Establishing SSL connection without server's identity verification is not recommende
11:47:07,357 DEBUG SpringManagedTransaction:87 - JDBC Connection [com.mysql.cj.jdbc.ConnectionImpl@79a1728c] will not b
11:47:07,437 DEBUG findById:159 - ==>  Preparing: SELECT * FROM ay_user WHERE id = ?
11:47:07,879 DEBUG findById:159 - ==> Parameters: 1(String)                          第一次查询日志
11:47:08,223 DEBUG findById:159 - <==      Total: 1
name: ay  password:123
name: ay  password:123  第二次查询
11:47:19,041 DEBUG DataSourceUtils:340 - Returning JDBC Connection to DataSource
```

图 7-2　控制台打印的信息

由图 7-2 中控制台打印的信息可以看出第一次查询和第二次查询的结果，查询 SQL 的日志只输出了一遍，这就说明了第二次查询的数据不是从数据库查询出来的，而是从一级缓存中获取的。

现在在两次查询之间执行 commit 操作（更新、删除或者插入），具体代码如下：

```
@Resource
private SqlSessionFactoryBean sqlSessionFactoryBean;

@Test
public void testSessionCache() throws Exception{
    SqlSessionFactory sqlSessionFactory = sqlSessionFactoryBean.getObject();
    SqlSession sqlSession = sqlSessionFactory.openSession();
    AyUserDao ayUserDao = sqlSession.getMapper(AyUserDao.class);
    //第一次查询
    AyUser ayUser = ayUserDao.findById("1");
    System.out.println("name: " + ayUser.getName()
                + "  password:" + ayUser.getPassword());

    //执行 commit 操作（如更新、插入、删除等操作）
    AyUser user = new AyUser();
    user.setId(1);
    user.setName("al");
    ayUserDao.update(new AyUser());
    ayUserDao.update(ayUser);

    //第二次查询
    AyUser ayUser2 = ayUserDao.findById("1");
    System.out.println("name: " + ayUser2.getName()
            + "  password:" + ayUser2.getPassword());
    sqlSession.close();;
}
```

执行测试用例 testSessionCache()，控制台打印相关的信息，具体如图 7-3 所示。

```
11:52:29,402 DEBUG DataSourceUtils:114 - Fetching JDBC Connection from DataSource
Thu May 24 11:52:29 CST 2018 WARN: Establishing SSL connection without server's identity verification is not recommen
11:52:29,722 DEBUG SpringManagedTransaction:87 - JDBC Connection [com.mysql.cj.jdbc.ConnectionImpl@71391b3f] will not
11:52:29,730 DEBUG findById:159 - ==>  Preparing: SELECT * FROM ay_user WHERE id = ?    第一次查询日志
11:52:29,795 DEBUG findById:159 - ==> Parameters: 1(String)
11:52:29,855 DEBUG findById:159 - <==      Total: 1
name: ay  password:123
11:52:29,857 DEBUG update:159 - ==>  Preparing: UPDATE ay_user SET name = ?, password = ? WHERE id = ?    更新日志
11:52:29,858 DEBUG update:159 - ==> Parameters: ay(String), 123(String), 1(Integer)
11:52:29,866 DEBUG update:159 - <==    Updates: 1
11:52:29,867 DEBUG findById:159 - ==>  Preparing: SELECT * FROM ay_user WHERE id = ?    第二次查询日志
11:52:29,868 DEBUG findById:159 - ==> Parameters: 1(String)
11:52:29,871 DEBUG findById:159 - <==      Total: 1
name: ay  password:123
11:52:29,872 DEBUG DataSourceUtils:340 - Returning JDBC Connection to DataSource
```

图 7-3　控制台打印的信息

由图 7-3 控制台打印的信息可以看出，第一次查询和第二次查询之间执行了 update 操作，update 操作会执行 commit，也就是说会清空一级缓存来保证数据的最新状态，防止脏读情况出现。因此第一次查询 id 为 1 的用户信息时，预编译了 SQL 语句，数据从数据库中查询出结果；第二次查询之前更新了用户数据，并执行了 commit 方法提交了修改，没有在一级缓存中找到 id 为 1 的用户信息，所以再次通过数据库进行查询。

7.2.3　一级缓存生命周期

MyBatis 在开启一个 Session 会话时，会创建一个新的 SqlSession 对象，每个 SqlSession 对象会创建一个新的 Executor 对象，Executor 对象中持有一个新的 PerpetualCache 对象；当会话结束时，SqlSession 对象及其内部的 Executor 对象还有 PerpetualCache 对象也会一并释放掉，即：

（1）如果 SqlSession 调用了 close()方法，会释放掉一级缓存 PerpetualCache 对象，一级缓存不可用。

（2）如果 SqlSession 调用了 clearCache()，会清空 PerpetualCache 对象中的数据，但是该对象仍可使用。

（3）SqlSession 中执行任何一个 DML 操作（update、delete、insert），都会清空 PerpetualCache 对象的数据，但是该对象可以继续使用。

7.3　二级查询缓存

7.3.1　二级缓存概述

二级缓存是 Mapper 级别的缓存。使用二级缓存时，多个 SqlSession 使用同一个 Mapper（namespace）的 SQL 语句操作数据库，得到的数据会存在二级缓存区域，二级缓存同样是使用 HashMap 进行数据存储。二级缓存比一级缓存作用域范围更大，多个 SqlSession 可以共用二级缓存，二级缓存是跨 SqlSession 的。当某个 SqlSession 类的实例对象执行了增、删、改等操作时，Mapper 实例会清空二级缓存。MyBatis 默认没有开启二级缓存，需要在配置中

开启二级缓存。开启二级缓存的步骤如下：

首先，在applicationContext.xml配置文件中添加如下配置：

```
    <!--3. 配置SqlSessionFactory对象-->
    <bean id="sqlSessionFactory"
class="org.mybatis.spring.SqlSessionFactoryBean">
        <!--注入数据库连接池-->
        <property name="dataSource" ref="dataSource"/>
        <!--扫描sql配置文件:mapper需要的xml文件-->
        <property name="mapperLocations" value="classpath:mapper/*.xml"/>
        <!-- mybatis配置文件的位置 -->
        <property name="configLocation" value="classpath:mybatis-
config.xml"></property>
    </bean>
```

上面配置信息中最重要的就是指定MyBatis配置文件的位置，即

```
    <!-- mybatis配置文件的位置 -->
    <property name="configLocation" value="classpath:mybatis-
config.xml"></property>
```

其次，在src\main\resources目录下添加配置文件mybatis-config.xml，具体代码如下：

```
    <?xml version="1.0" encoding="UTF-8" ?>
    <!DOCTYPE configuration
            PUBLIC "-//mybatis.org//DTD Config 3.0//EN"
            "http://mybatis.org/dtd/mybatis-3-config.dtd">
    <configuration>

        <!-- 全局配置参数，需要时再设置 -->
        <settings>
            <!-- 开启二级缓存  默认是不开启的-->
            <setting name="cacheEnabled" value="true"/>
        </settings>

    </configuration>
```

最后，由于二级缓存是Mapper级别的，还要在需要开启二级缓存的具体mapper.xml文件中开启二级缓存，方法很简单，只需要在mapper.xml文件中添加一个cache标签既可，具体代码如下：

```
    <!-- 开启AyUserMapper的namespace下的二级缓存 -->
    <cache/>
```

cache标签有很多属性，常用的属性如表7-1所示。

表 7-1　cache 标签的属性

属性名称	描　述
eviction	收回策略，默认为 LRU。有如下几种： • LRU（最近最少使用的策略）：移除最长时间不被使用的对象 • FIFO（先进先出策略）：按对象进入缓存的顺序来移除它们 • SOFT（软引用策略）：移除基于垃圾回收器状态和软引用规则的对象 • WEAK（弱引用策略）：更积极地移除基于垃圾收集器状态和弱引用规则的对象
flushInterval	刷新间隔，可以被设置为任意的正整数，而且它们代表一个合理的毫秒形式的时间段。默认情况是不设置，也就是没有刷新间隔，缓存仅仅调用语句时刷新
readOnly	只读，属性可以被设置为 true 或 false。只读的缓存会给所有调用者返回缓存对象的相同实例，因此这些对象不能被修改。这提供了很重要的性能优势，可读写的缓存会返回缓存对象的拷贝（通过序列化）。这会慢一些，但是安全，因此默认是 false
size	缓存数目，可以被设置为任意正整数，要记住你缓存的对象数目和你运行环境的可用内存资源数目。默认值是 1024

7.3.2　二级缓存示例

下面来看 MyBatis 二级缓存的应用实例，具体代码如下：

```
@Resource
private SqlSessionFactoryBean sqlSessionFactoryBean;

@Test
public void testSessionCache() throws Exception{
    SqlSessionFactory sqlSessionFactory = sqlSessionFactoryBean.getObject();
    SqlSession sqlSession = sqlSessionFactory.openSession();
    AyUserDao ayUserDao = sqlSession.getMapper(AyUserDao.class);
    //第一次查询
    AyUser ayUser = ayUserDao.findById("1");
    System.out.println("name: " + ayUser.getName()
                    + "  password:" + ayUser.getPassword());

    //执行 commit 操作（如：更新、插入、删除等操作）
    AyUser user = new AyUser();
    user.setId(1);
    user.setName("al");
    ayUserDao.update(new AyUser());
    ayUserDao.update(ayUser);

    //第二次查询（命中缓存）
    AyUser ayUser2 = ayUserDao.findById("1");
```

```
    System.out.println("name: " + ayUser2.getName()
            + "  password:" + ayUser2.getPassword());
    sqlSession.close();;
}
```

AyUserDao 和 AyUserMapper.xml 代码如下：

```
@Repository
public interface AyUserDao {

    AyUser findById(String id);
}

<select id="findById"  parameterType="String" resultMap="userMap">
    SELECT * FROM ay_user
    WHERE id = #{id}
</select>
```

当开启 MyBatis 二级缓存后，执行测试用例 testSessionCache()，控制台打印相关的信息，具体如图 7-4 所示。

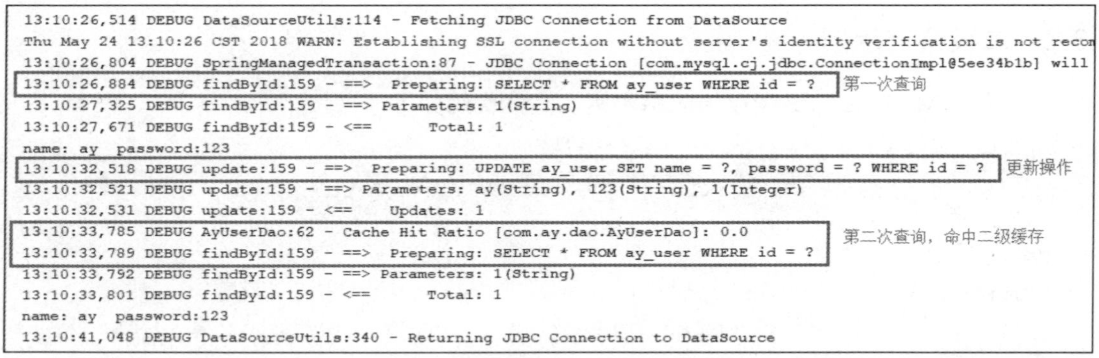

```
13:10:26,514 DEBUG DataSourceUtils:114 - Fetching JDBC Connection from DataSource
Thu May 24 13:10:26 CST 2018 WARN: Establishing SSL connection without server's identity verification is not recon
13:10:26,804 DEBUG SpringManagedTransaction:87 - JDBC Connection [com.mysql.cj.jdbc.ConnectionImpl@5ee34b1b] will
13:10:26,884 DEBUG findById:159 - ==>  Preparing: SELECT * FROM ay_user WHERE id = ?    第一次查询
13:10:27,325 DEBUG findById:159 - ==> Parameters: 1(String)
13:10:27,671 DEBUG findById:159 - <==      Total: 1
name: ay  password:123
13:10:32,518 DEBUG update:159 - ==>  Preparing: UPDATE ay_user SET name = ?, password = ? WHERE id = ?    更新操作
13:10:32,521 DEBUG update:159 - ==> Parameters: ay(String), 123(String), 1(Integer)
13:10:32,531 DEBUG update:159 - <==    Updates: 1
13:10:33,785 DEBUG AyUserDao:62 - Cache Hit Ratio [com.ay.dao.AyUserDao]: 0.0    第二次查询，命中二级缓存
13:10:33,789 DEBUG findById:159 - ==>  Preparing: SELECT * FROM ay_user WHERE id = ?
13:10:33,792 DEBUG findById:159 - ==> Parameters: 1(String)
13:10:33,801 DEBUG findById:159 - <==      Total: 1
name: ay  password:123
13:10:41,048 DEBUG DataSourceUtils:340 - Returning JDBC Connection to DataSource
```

图 7-4　控制台打印的信息

由图 7-4 可知，第一次查询数据时，获取连接、编译 SQL、加载了数据库中的数据。而第二次查询数据之前，进行了 update 操作，相当于进行 commit 操作，也就是说会清空一级缓存来保证数据的最新状态。但是开启了二级缓存，在第二次查询时，会从二级缓存中获取数据。

这里需要注意的是，如果在 select 标签中设置“userCache = false”可以禁用当前 select 语句的二级缓存，具体代码如下：

```
<select id="findById" useCache="false"  parameterType="String"
resultMap="userMap">
    SELECT * FROM ay_user
    WHERE id = #{id}
</select>
```

这里简单总结一下二级缓存的特点：

- 缓存是以namespace为单位的，不同的namespace下的操作互不影响。
- 增删改查操作会清空namespace下的全部缓存。

还需要注意的是，使用二级缓存需要特别谨慎，有时候不同的 namespace 下的 SQL 配置可能缓存了相同的数据。例如 AyUserMapper.xml 中有很多查询缓存了用户数据，其他的 XXXMapper.xml 中有针对用户表进行单表操作，也缓存了用户数据，如果在 AyUserMapper.xml 中做了刷新缓存的操作，在 XXXMapper.xml 中的缓存数据仍然有效，这样在查询数据时可能会出现脏数据。所以使用 MyBatis 的二级缓存时，要根据具体的业务情况谨慎使用。

7.3.3 Cache-ref 共享缓存

MyBatis 并不是整个 Application 只有一个 Cache 缓存对象，它将缓存划分的更细，也就是 Mapper 级别的，即每一个 Mapper 都可以拥有一个 Cache 对象，具体如下：

（1）为每一个 Mapper 分配一个 Cache 缓存对象（使用<cache>节点配置）。

（2）多个 Mapper 共用一个 Cache 缓存对象（使用<cache-ref>节点配置）。

如果想让多个 Mapper 共用一个 Cache，可以使用<cache-ref namespace="">节点来指定这个 Mapper 共享哪一个 Mapper 的 Cache 缓存，具体如图 7-5 所示。

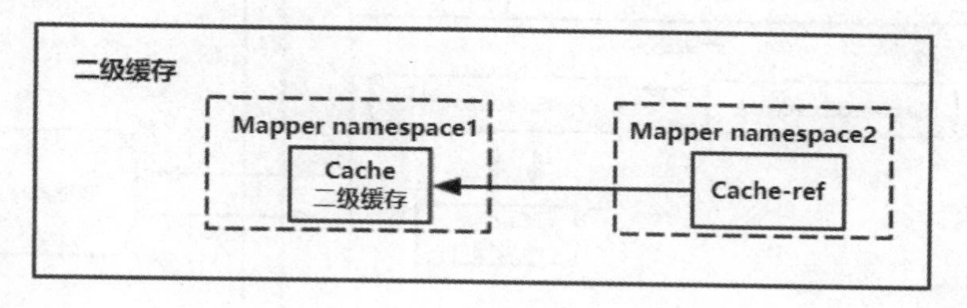

图 7-5　控制台打印的信息

<cache-ref>标签的使用示例如下。

UserMapper.xml 代码如下：

```
<?xml version="1.0" encoding="UTF-8" ?>
<!DOCTYPE mapper PUBLIC "-//mybatis.org//DTD Mapper 3.0//EN"
      "http://mybatis.org/dtd/mybatis-3-mapper.dtd">
<mapper namespace="com.ay.dao.UserDao">
   <!-- 非常重要 -->
    <cache/>
    //省略代码
</mapper>
```

MoodMapper.xml 代码如下：

```
<?xml version="1.0" encoding="UTF-8" ?>
<!DOCTYPE mapper PUBLIC "-//mybatis.org//DTD Mapper 3.0//EN"
```

```
        "http://mybatis.org/dtd/mybatis-3-mapper.dtd">
<mapper namespace="com.ay.dao.MoodDao">
    //共享 UserMapper 的二级缓存，要求 UserMapper.xml 必须有<cache/>标签
    <cache-ref namespace="com.ay.dao.UserDao"/>
    //省略代码
</mapper>
```

7.4 MyBatis 缓存原理

7.4.1 MyBatis 缓存的工作机制

如图 7-6 所示，一个 SqlSession 对象中创建一个本地缓存（local cache），对于每次查询都会根据查询条件去一级缓存中查找，如果缓存中存在数据，就直接从缓存中取出，然后返回给用户；否则，从数据库读取数据，将查询结果存入缓存并返回给用户。

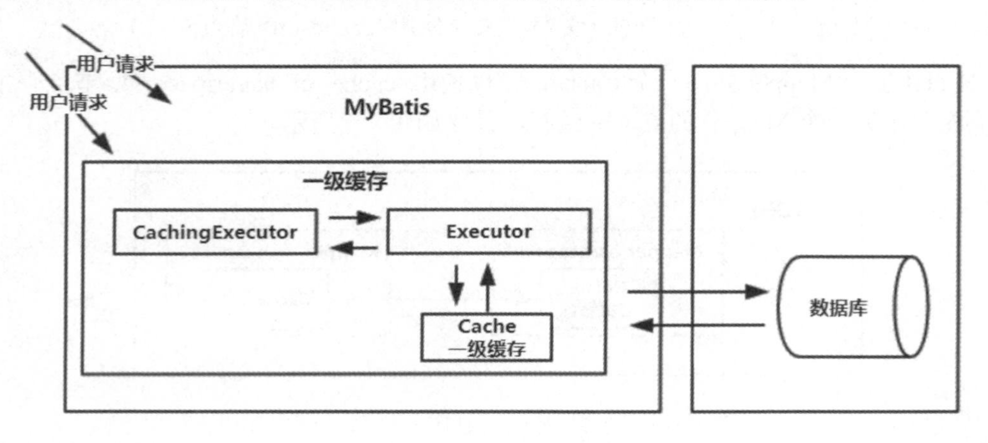

图 7-6　MyBatis 一级缓存机制

SqlSession 将它的工作交给了 Executor 执行器这个角色来完成，负责完成对数据库的各种操作。当创建一个 SqlSession 对象时，MyBatis 会为这个 SqlSession 对象创建一个新的 Executor 执行器，而缓存信息就被维护在这个 Executor 执行器中，MyBatis 将缓存和对缓存相关的操作封装成了 Cache 接口。

如图 7-7 所示，MyBatis 的二级缓存机制的关键是使用 Executor 对象。当开启 SqlSession 会话时，一个 SqlSession 对象使用一个 Executor 对象来完成会话操作。如果用户配置了 "cacheEnabled=true"，那么 MyBatis 在为 SqlSession 对象创建 Executor 对象时，会对 Executor 对象加上一个装饰者：CachingExecutor，这时 SqlSession 使用 CachingExecutor 对象来完成操作请求。CachingExecutor 对于查询请求会先判断该查询请求在二级缓存中是否有缓存结果，如果有查询结果，则直接返回缓存结果；如果缓存中没有，再交给真正的 Executor 对象来完成查询操作，之后 CachingExecutor 会将 Executor 返回的查询结果放置到缓存中，然后再返回给用户。

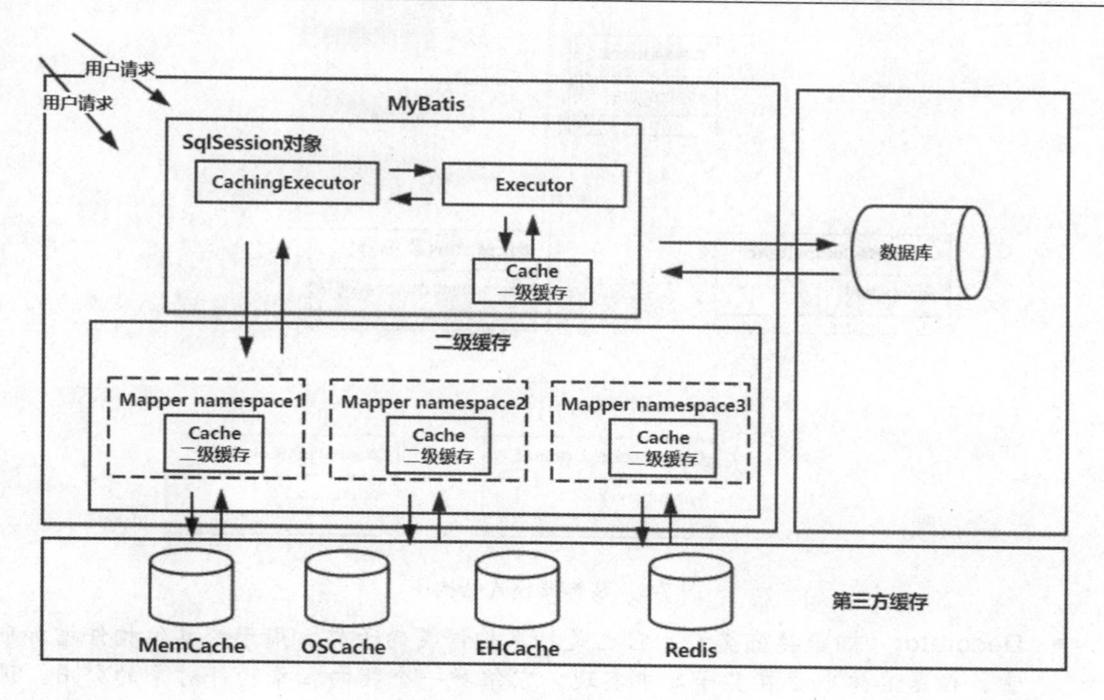

图 7-7　MyBatis 二级缓存机制

7.4.2　装饰器模式

装饰器模式（Decorator Pattern）可以在不改变一个对象本身功能的基础上给对象增加额外的功能。装饰器模式是一种用于替代继承的技术，它通过一种无须定义子类的方式来给对象动态增加职责，使用对象之间的关联关系取代类之间的继承关系。在装饰器模式中引入了装饰类，在装饰类中既可以调用待装饰的原有类的方法，还可以增加新的方法，以扩充原有类的功能。

装饰器模式动态地给一个对象增加一些额外的职责，就增加对象功能来说，装饰器模式比生成子类实现更为灵活。装饰器模式是一种对象结构型模式。

在装饰器模式中，为了让系统具有更好的灵活性和可扩展性，通常会定义一个抽象装饰类，而将具体的装饰类作为它的子类，装饰器模式的结构如图 7-8 所示。

在装饰器模式结构图中包含了如下几个角色：

- Component（抽象构件）：它是具体构件和抽象装饰类的共同父类，声明了在具体构件中实现的业务方法，它的引入可以使客户端以一致的方式处理未被装饰的对象以及装饰之后的对象，实现客户端的透明操作。
- ConcreteComponent（具体构件）：它是抽象构件类的子类，用于定义具体的构件对象，实现了在抽象构件中声明的方法，装饰器可以给它增加额外的职责（方法）。

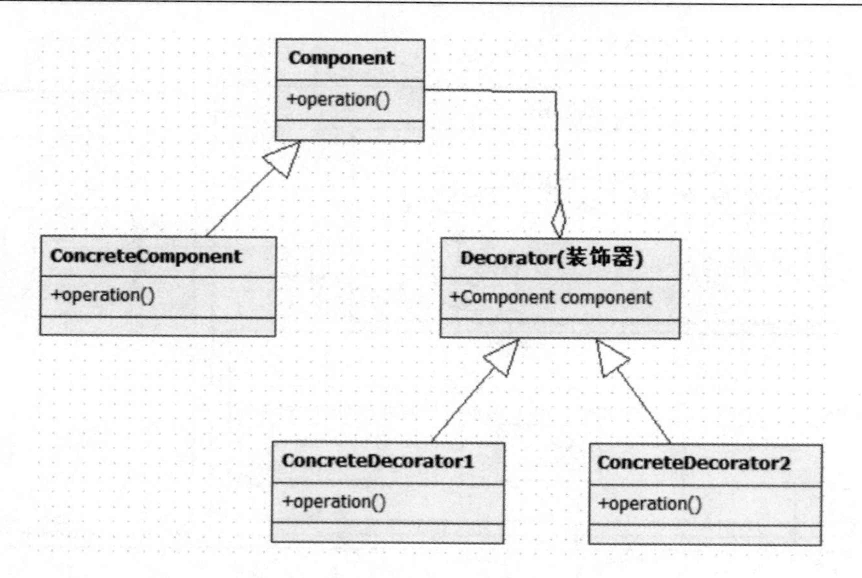

图 7-8　装饰器模式结构图

- Decorator（**抽象装饰类**）：它也是抽象构件类的子类，用于给具体构件增加职责，但是具体职责在其子类中实现。它维护一个指向抽象构件对象的引用，通过该引用可以调用装饰之前构件对象的方法，并通过其子类扩展该方法，以达到装饰的目的。
- ConcreteDecorator（**具体装饰类**）：它是抽象装饰类的子类，负责向构件添加新的职责。每一个具体装饰类都定义了一些新的行为，它可以调用在抽象装饰类中定义的方法，并可以增加新的方法用以扩充对象的行为。

由于具体构件类和装饰类都实现了相同的抽象构件接口，因此装饰器模式以对客户透明的方式动态地给一个对象附加上更多的责任，换言之，客户端并不会觉得对象在装饰前和装饰后有什么不同。装饰器模式可以在不需要创造更多子类的情况下，将对象的功能加以扩展。

装饰器模式的核心在于抽象装饰类的设计，Decorator（装饰器）的典型代码如下：

```
class Decorator implements Component{
    //维持一个对抽象构件对象的引用
    private Component component;
    //注入一个抽象构件类型的对象
    public Decorator(Component component){
        this.component = component;
    }
    //调用原有业务方法
    public void operation(){
        component.operation();
    }
}
```

在抽象装饰类 Decorator 中定义 Component 类型的对象，维持一个对抽象构件对象的引用，并可以通过构造方法或 Setter 方法将一个 Component 类型的对象注入进来，同时由于 Decorator 类实现了抽象构件 Component 接口，因此需要实现在其中声明的业务方法 operation()。需要注意的是，在 Decorator 中并未真正实现 operation()方法，而只是调用原有 component 对象的 operation()方法，它没有真正实施装饰，而是提供一个统一的接口，将具体装饰过程交给子类完成。

Decorator 的子类即具体装饰类 ConcreteDecorator 中将继承 operation()方法并根据需要进行扩展，典型的具体装饰类代码如下：

```
class ConcreteDecorator extends Decorator{

    public ConcreteDecorator(Component  component){
        super(component);
    }
    public void operation(){
        //调用原有业务方法
        super.operation();
        //调用新增业务方法
        addedBehavior();
    }
    //新增业务方法
    public  void addedBehavior(){

    }
}
```

7.4.3　Cache 接口及其实现

Cache 接口是 MyBatis 缓存模块中最核心的接口，它定义了所有缓存的基本行为。Cache 接口的具体源码如下：

```
public interface Cache {
    //该缓存对象的 id
    String getId();
    //项缓存添加数据，一般情况下 key 为 CacheKey，value 为查询结果
    void putObject(Object key, Object value);
    //根据指定的 key 在缓存中查找对应的结果对象
    Object getObject(Object key);
    //删除 key 对应的缓存项
    Object removeObject(Object key);
    //清空缓存
    void clear();
    //缓存项个数
    int getSize();
    //获取读写锁
```

```
    ReadWriteLock getReadWriteLock();

}
```

Cache 接口的实现类有很多，具体如图 7-9 所示。

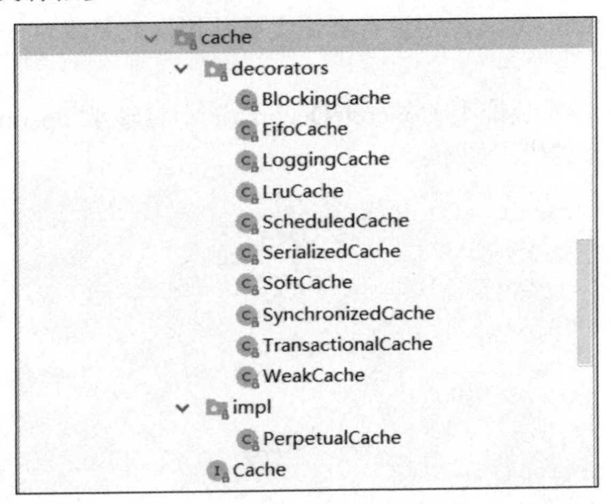

图 7-9 Cache 接口的实现类

在 Cache 接口的实现类中，大部分都是装饰器，只有 PerpetualCache 提供了 Cache 接口的基本实现。PerpetualCache 在缓存模块中扮演着 ConcreteComponent（具体构件）的角色，底层使用 HashMap 记录缓存项，PerpetualCache 具体的源码如下：

```
public class PerpetualCache implements Cache {

  private final String id;

  private Map<Object, Object> cache = new HashMap<Object, Object>();

  public PerpetualCache(String id) {}

  @Override
  public String getId() {}

  @Override
  public int getSize() {}

  @Override
  public void putObject(Object key, Object value) {}

  @Override
  public Object getObject(Object key) {}

  @Override
```

```
    public Object removeObject(Object key) {}

    @Override
    public void clear() {}

    @Override
    public ReadWriteLock getReadWriteLock() {}
}
```

除了 PerpetualCache 缓存类外，Cache 接口的其他实现类都是装饰器，这些装饰器扮演着 ConcreteDecorator 的角色并在 PerpetualCache 的基础上提供额外的功能，通过多个组合后满足一个特定的需求。其他装饰器和 Cache 的类结构如图 7-10 所示。

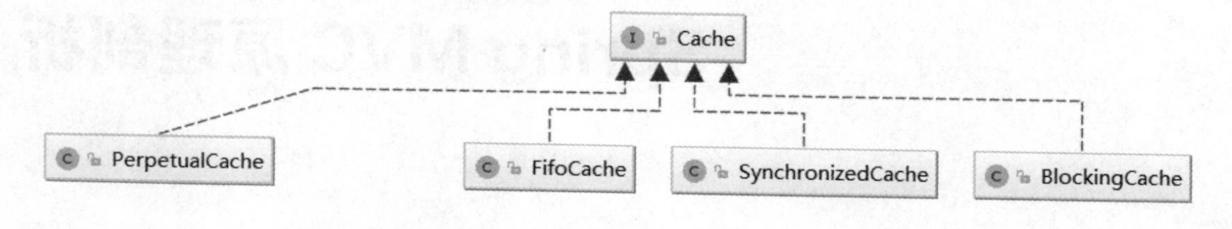

图 7-10 Cache 接口的实现类

这里就不再继续剖析每个装饰器缓存类的源码，感兴趣的读者可以自己查看 MyBatis 相关源码进行学习。

7.5 思考与练习

1. 简述 MyBatis 一级缓存和二级缓存。
2. 默认情况下，MyBatis 是否只开启一级缓存？
3. 动手实践 MyBatis 一级缓存。
4. 简述 MyBatis 一级缓存的生命周期。
5. 简述一级缓存和二级缓存的区别。
6. 简单描述 MyBatis 一级缓存的工作原理。
7. 简单描述 MyBatis 二级缓存的工作原理。
8. 简述 MyBatis 的 Cache 接口以及接口包含的方法。
9. Cache 接口提供哪些实现类？

第 8 章 Spring MVC 原理剖析

本章主要介绍 Spring MVC 执行流程的原理、前端控制器 DispatcherServlet 的原理、处理映射器和处理适配器的原理以及视图解析器的原理等。

8.1 Spring MVC 的执行流程与前端控制器

Spring MVC 框架整体的请求流程如图 8-1 所示，该图显示了用户从请求到响应的完整流程。

（1）用户发起 request 请求，该请求被前端控制器（DispatcherServlet）处理。

（2）前端控制器（DispatcherServlet）请求处理映射器（HandlerMapping）查找 Handler。

（3）处理映射器（HandlerMapping）根据配置查找相关的 Handler，返回给前端控制器（DispatcherServlet）。

（4）前端控制器（DispatcherServlet）请求处理适配器（HandlerAdapter）执行相应的 Handler（或称为 Controller）。

（5）处理适配器（HandlerAdapter）执行 Handler。

（6）Handler 执行完毕后会返回 ModelAndView 对象给 HandlerAdapter。

（7）HandlerAdapter 对象接收到 Handler 返回的 ModelAndView 对象后，将其返回给前端控制器（DispatcherServlet）。

（8）前端控制器（DispatcherServlet）接收到 ModelAndView 对象后，请求视图解析器（View Resolver）对视图进行解析。

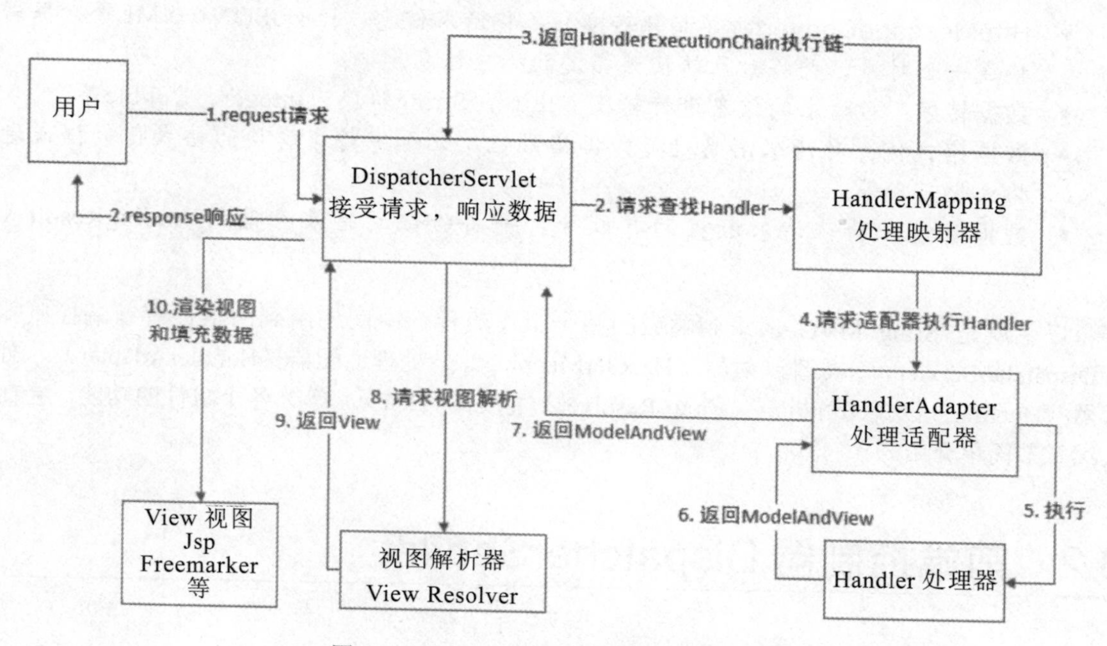

图 8-1 Spring MVC 框架整体的请求流程

（9）视图解析器（View Resolver）根据 View 信息匹配相应的视图结果，返回给前端控制器（DispatcherServlet）。

（10）前端控制器（DispatcherServlet）收到 View 视图后，对视图进行渲染，将 Model 中的模型数据填充到 View 视图中的 request 域，生成最终的视图。

（11）前端控制器（DispatcherServlet）返回请求结果给用户。

处理适配器（HandlerAdapter）执行 Handler（或称为 Controller）的过程中，Spring 还做了一些额外的工作，具体如图 8-2 所示。

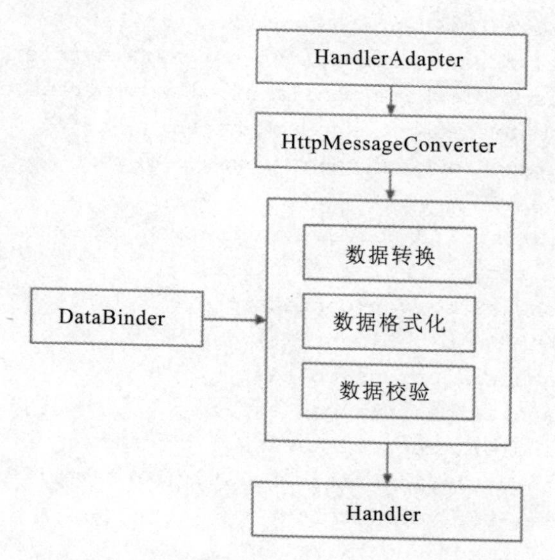

图 8-2 数据转换、格式化、校验

- **HttpMessageConverter（消息转换）**：将请求信息，比如JSON、XML等数据转换成一个对象，并将对象转换为指定的响应信息。
- **数据转换**：对请求的信息进行转换，比如，String转换为Integer、Double等。
- **数据格式化**：对请求消息进行数据格式化，比如字符串转换为格式化数据或者格式化日期等。
- **数据验证**：验证请求数据的有效性，并将验证的结果存储到BindingResult或Error中。

以上就是 Sring MVC 请求到响应的整个工作流程，中间使用到的组件有前端控制器（DispatcherServlet）、处理映射器（HandlerMapping）、处理适配器（HandlerAdapter）、处理器（Handler）、视图解析器（View Resolver）和视图（View）等。各个组件的功能，会在后续章节简单介绍。

8.2 前端控制器 DispatcherServlet

前端控制器 DispatcherServlet 的作用就是接受用户请求，然后给用户响应结果。它的作用相当于一个转发器或中央处理器，控制整个流程的执行，对各个组件进行统一调度，以降低组件之间的耦合性，有利于组件之间的扩展。

DispatcherServlet 的部分源码如下所示：

```
public class DispatcherServlet extends FrameworkServlet {

    private LocaleResolver localeResolver;
    private ThemeResolver themeResolver;
        private List<HandlerMapping> handlerMappings;
    private List<HandlerAdapter> handlerAdapters;
        private List<HandlerExceptionResolver> handlerExceptionResolvers;
    private RequestToViewNameTranslator viewNameTranslator;
    private FlashMapManager flashMapManager;
        private List<ViewResolver> viewResolvers;
    //省略代码
    protected void initStrategies(ApplicationContext context) {
            initMultipartResolver(context);
            initLocaleResolver(context);
            initThemeResolver(context);
            initHandlerMappings(context);
            initHandlerAdapters(context);
            initHandlerExceptionResolvers(context);
            initRequestToViewNameTranslator(context);
            initViewResolvers(context);
            initFlashMapManager(context);
    }
```

```
}
```

DispatcherServlet 类的类继承结构如图 8-3 所示。

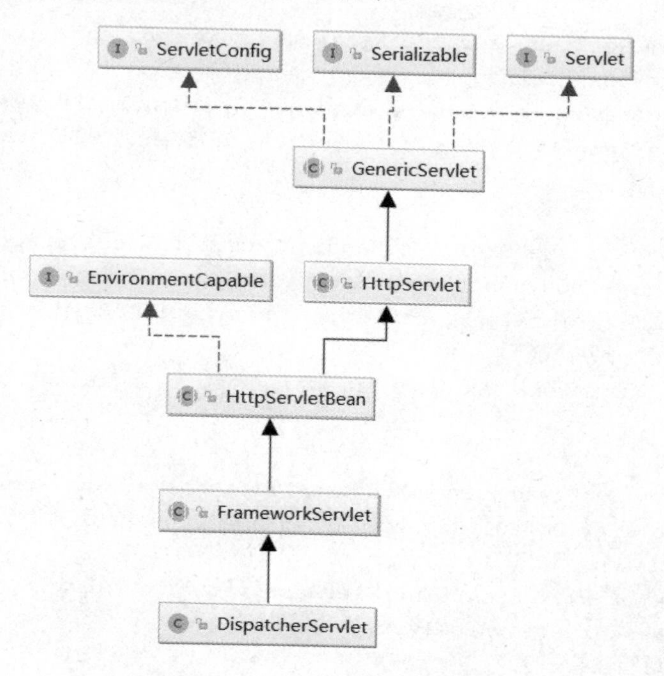

图 8-3　DispatcherServlet 的类结构

由图 10-3 可知，DispatcherServlet 最上层的父类是 Servlet 类，也就是说 DispatcherServlet 也是一个 Servlet，且包含有 deGet()和 doPost()方法。initStrategies 方法在 WebApplicationContext 初始化后自动执行，自动扫描上下文的 Bean，根据名称或者类型匹配的机制查找自定义的组件，如果没有找到，会装配 Spring 的默认组件。Spring 的默认组件在 org.springframework.web.servlet 路径下的 DispatcherServlet.properties 配置文件中配置。DispatcherServlet.properties 的具体代码如下：

```
    # Default implementation classes for DispatcherServlet's strategy
interfaces.
    # Used as fallback when no matching beans are found in the
DispatcherServlet context.
    # Not meant to be customized by application developers.
    //本地化解析器
    org.springframework.web.servlet.LocaleResolver=org.springframework.web.ser
vlet.i18n.AcceptHeaderLocaleResolver
    //主题解析器
    org.springframework.web.servlet.ThemeResolver=org.springframework.web.serv
let.theme.FixedThemeResolver
    //处理映射器
    org.springframework.web.servlet.HandlerMapping=org.springframework.web.ser
vlet.handler.BeanNameUrlHandlerMapping,\
```

```
    org.springframework.web.servlet.mvc.method.annotation.RequestMappingHandle
rMapping
    //处理适配器
    org.springframework.web.servlet.HandlerAdapter=org.springframework.web.ser
vlet.mvc.HttpRequestHandlerAdapter,\
        org.springframework.web.servlet.mvc.SimpleControllerHandlerAdapter,\
        org.springframework.web.servlet.mvc.method.annotation.RequestMappingHa
ndlerAdapter
    //异常处理器
    org.springframework.web.servlet.HandlerExceptionResolver=org.springframewo
rk.web.servlet.mvc.method.annotation.ExceptionHandlerExceptionResolver,\
        org.springframework.web.servlet.mvc.annotation.ResponseStatusException
Resolver,\
        org.springframework.web.servlet.mvc.support.DefaultHandlerExceptionRes
olver
    //视图名称解析器
    org.springframework.web.servlet.RequestToViewNameTranslator=org.springfram
ework.web.servlet.view.DefaultRequestToViewNameTranslator
    //视图解析器
    org.springframework.web.servlet.ViewResolver=org.springframework.web.servl
et.view.InternalResourceViewResolver
    //FlashMap 映射管理器
    org.springframework.web.servlet.FlashMapManager=org.springframework.web.se
rvlet.support.SessionFlashMapManager
```

DispatcherServlet 类包含许多方法，大致可以分为以下三类：

（1）初始化相关处理类的方法，比如 initMultipartResolver()、initLocaleResolver()等。
（2）响应 HTTP 请求的方法。
（3）执行处理请求逻辑的方法。

DispatcherServlet 装配的组件，具体内容如下所示：

- **本地化解析器（LocaleResolver）**：本地化解析，只允许一个实例。因为Spring支持国际化，所以LocalResover解析客户端的Locale信息从而方便进行国际化。如果没有找到，使用默认的实现类AcceptHeaderLocaleResolver作为该类型的组件。
- **主题解析器（ThemeResovler）**：主题解析，只允许一个实例。通过它来实现一个页面多套风格，即常见的类似于软件皮肤效果。如果没有找到，使用默认的实现类FixedThemeResolver作为该类型的组件。
- **处理映射器（HandlerMapping）**：请求到处理器的映射，允许多个实例。如果映射成功返回一个HandlerExecutionChain对象（包含一个Handler处理器[页面控制器]）对象、多个HandlerInterceptor拦截器）对象；如果detectHandlerMappings的属性为true（默认为true），则根据类型匹配机制查找上下文及Spring容器中所有类型为HandlerMapping的Bean，将它们作为该类型的组件。如果detectHandlerMappings的属性为false，则查找名为handlerMapping、类型为HandlerMapping的Bean作为该类型组件。如果以上两种方式都没有找到，则使用BeanNameUrlHandlerMapping实现类创建该类型的组件。BeanNameUrlHandlerMapping将URL与Bean名字映射，映射成功的Bean就是此处的处理器。

- **处理适配器（HandlerAdapter）**：允许多个实例，HandlerAdapter将会把处理器包装为适配器，从而支持多种类型的处理器，即适配器设计模式的应用，从而很容易支持很多类型的处理器。如SimpleControllerHandlerAdapter将对实现了Controller接口的Bean进行适配，并且按处理器的handleRequest方法进行功能处理。默认使用DispatcherServlet.properties配置文件中指定的三个实现类分别创建一个适配器，并将其添加到适配器列表中。
- **处理异常解析器（HandlerExceptionResolver）**：允许多个实例。处理器异常解析可以将异常映射到相应的统一错误界面，从而显示用户友好的界面（而不是给用户看到具体的错误信息）。默认使用DispatcherServlet.properties配置文件中定义的实现类。
- **视图名称解析器（ViewNameTranslator）**：只允许一个实例。默认使用DefaultRequestToViewNameTranslator作为该类型的组件。
- **视图解析器（ViewResolver）**：允许多个实例。ViewResolver将把逻辑视图名解析为具体的View，通过这种策略模式，很容易更换其他视图技术，如InternalResourceViewResolver将逻辑视图名映射为JSP视图。
- **FlashMap映射管理器**：查找名为FlashMapManager、类型为SessionFlashMap-Manager的bean作为该类型组件，用于管理FlashMap，即数据默认存储在HttpSession中。

需要注意的是，DispatcherServlet 装配的各种组件，有些只允许一个实例，有些则允许多个实例。如果同一个类型的组件存在多个，可以通过 Order 属性确定优先级的顺序，值越小的优先级越高。

8.3 处理映射器和适配器

8.3.1 处理映射器

处理映射器 HandlerMapping 是指请求到处理器的映射时，允许有多个实例。如果映射成功返回一个 HandlerExecutionChain 对象（包含一个 Handler 处理器[页面控制器]对象、多个 HandlerInterceptor 拦截器对象）。Spring MVC 提供了多个处理映射器 HandlerMapping 实现类，下面分别进行说明。

1. BeanNameUrlHandlerMapping

BeanNameUrlHandlerMapping 是默认映射器，在不配置的情况下，默认就使用这个类来映射请求。其映射规则是根据请求的 URL 与 Spring 容器中定义的处理器 bean 的 name 属性值进行匹配，从而在 Spring 容器中找到 Handler（处理器）的 bean 实例。

```
//默认映射器，在不配置的情况下，默认就使用这个来映射请求
<bean class="org.springframework.web.servlet.
handler.BeanNameUrlHandlerMapping"></bean>
//映射器把请求映射到 controller
<beanid="testController" name="/hello.do"
class="cn.itcast.controller.TestController"></bean>
```

2. SimpleUrlHandlerMapping

SimpleUrlHandlerMapping 根据浏览器 URL 匹配 prop 标签中的 key，通过 key 找到对应的 Controller。

```
    <bean class="org.springframework.web.servlet.handler.
SimpleUrlHandlerMapping">
        <property name="mappings">
            <props>
                <prop key="/hello.do">testController</prop>
                <prop key="/test.do">testController</prop>
            </props>
        </property>
    </bean>
    <bean id="testController"
    name="/hello.do" class="cn.itcast.controller.TestController"></bean>
```

上述配置了两个不同的 URL 映射，对应于同一个 Controller 配置，也就是说，在浏览器中发起两个不同的 URL 请求会得到相同的处理结果。

8.3.2 处理适配器

处理适配器（HandlerAdapter）允许多个实例，HandlerAdapter 将会把处理器包装为适配器，从而支持多种类型的处理器，即适配器设计模式的应用，从而很容易支持多种类型的处理器。如 SimpleControllerHandlerAdapter 将对实现了 Controller 接口的 Bean 进行适配，并且按处理器的 handleRequest 方法进行功能处理。默认使用 DispatcherServlet.properties 配置文件中指定的三个实现类分别创建一个适配器，并将其添加到适配器列表中。

Spring MVC 提供了多个处理适配器（HandlerAdapter）实现类，分别说明如下。

1. SimpleControllerHandlerAdapter

SimpleControllerHandlerAdapter 支持所有实现 Controller 接口的 Handler 控制器，是 Controller 实现类的适配器类，其本质是执行 Controller 类中的 handleRequest 方法。

SimpleControllerHandlerAdapter 的源码如下：

```
public class SimpleControllerHandlerAdapter implements HandlerAdapter {

        @Override
        public boolean supports(Object handler) {
        //判断是否实现 Controller 接口
            return (handler instanceof Controller);
        }

    @Override
        @Nullable
        public ModelAndView handle(HttpServletRequest request,
HttpServletResponse response, Object handler) throws Exception {
            //将 handler 强制转换为 Controller，并调用 handleRequest 方法
            return ((Controller) handler).handleRequest(request, response);
    }

    @Override
    public long getLastModified(HttpServletRequest request,
Object handler) {
            if (handler instanceof LastModified) {
                return ((LastModified) handler).getLastModified(request);
            }
            return -1L;
        }
}
```

Controller 接口的定义也很简单，仅仅定义了一个 handleRequest 方法，具体源码如下：

```
    @FunctionalInterface
public interface Controller {
    @Nullable
    ModelAndView handleRequest(HttpServletRequest request,
HttpServletResponse response) throws Exception;
}
```

2. HttpRequestHandlerAdapter

HttpRequestHandlerAdapter 本质是调用 HttpRequestHandler 的 handleRequest 方法，请看下述代码示例：

```
public class HttpRequestHandlerAdapter implements HandlerAdapter {
        @Override
        public boolean supports(Object handler) {
            //判断是否是 HttpRequestHandler 类型
return (handler instanceof HttpRequestHandler);
    }
```

```
        @Override
        @Nullable
        public ModelAndView handle(HttpServletRequest request,
    HttpServletResponse response, Object handler) throws Exception {
            //执行 HttpRequestHandler 的 handleRequest 方法
    ((HttpRequestHandler) handler).handleRequest(request, response);
            return null;
        }
        @Override
        public long getLastModified(HttpServletRequest request, Object
handler) {
        //返回 modified 值
            if (handler instanceof LastModified) {
                return ((LastModified) handler).getLastModified(request);
            }
            return -1L;
        }
    }
```

HttpRequestHandlerAdapter 本质是 HttpRequestHandler 的适配器，最终调用 HttpRequestHandler 的 handleRequest 方法。接口 HttpRequestHandler 的实现如下：

```
    @FunctionalInterface
    public interface HttpRequestHandler {
        oid handleRequest(HttpServletRequest request, HttpServletResponse
response)
            throws ServletException, IOException;
    }
```

3. RequestMappingHandlerAdapter

RequestMappingHandlerAdapter 其父类是 AbstractHandlerMethodAdapter 抽象类，AbstractHandlerMethodAdapter 只是简单地实现了 HandlerAdapter 中定义的接口，最终还是在 RequesrMappingHandlerAdapter 中对代码进行实现的，AbstractHandlerMethodAdapter 中增加了执行顺序 Order，具体如图 8-4 所示。

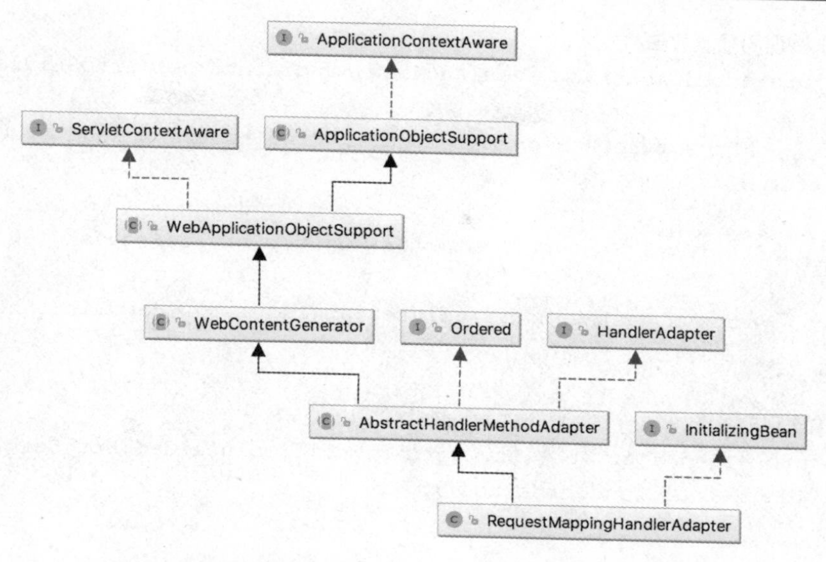

图 8-4　RequestMappingHandlerAdapter 类的继承关系

AbstractHandlerMethodAdapter 的源码如下：

```
public abstract class AbstractHandlerMethodAdapter
extends WebContentGenerator implements HandlerAdapter, Ordered {
        private int order = Ordered.LOWEST_PRECEDENCE;
        public AbstractHandlerMethodAdapter() {
            // no restriction of HTTP methods by default
            super(false);
        }
        public void setOrder(int order) {
            this.order = order;
        }

public int getOrder() {
            return this.order;
        }

        public final boolean supports(Object handler) {
            return (handler instanceof HandlerMethod &&
supportsInternal((HandlerMethod) handler));
        }
        protected abstract boolean supportsInternal(HandlerMethod
handlerMethod);

        public final ModelAndView handle(HttpServletRequest request,
HttpServletResponse response, Object handler) throws Exception {
            return handleInternal(request, response, (HandlerMethod)
handler);
        }
```

```
        //未实现的抽象方法
            protected abstract ModelAndView handleInternal(HttpServletRequest 
request,
                HttpServletResponse response, HandlerMethod handlerMethod) 
throws Exception;

            public final long getLastModified(HttpServletRequest request, 
Object handler) {
                return getLastModifiedInternal(request, (HandlerMethod) 
handler);
            }
        //未实现的抽象方法
            protected abstract long getLastModifiedInternal(HttpServletRequest 
request, HandlerMethod handlerMethod);
    }
```

RequestMappingHandlerAdapter 实现 AbstractHandlerMethodAdapter 类，真正意义上是实现了 HandlerAdapter 的功能。RequestMappingHandlerAdapter 的部分源码如下：

```
        public class RequestMappingHandlerAdapter extends 
AbstractHandlerMethodAdapter
            implements BeanFactoryAware, InitializingBean {
        //默认返回 true
        protected boolean supportsInternal(HandlerMethod handlerMethod) {
                return true;
            }
        //默认返回-1
    protected long getLastModifiedInternal(HttpServletRequest request, 
    HandlerMethod handlerMethod) {
                return -1;
            }
            //
        protected ModelAndView handleInternal(HttpServletRequest request,
                HttpServletResponse response, HandlerMethod handlerMethod) 
throws Exception {
                ModelAndView mav;
            //检查 request 请求方法 method 是否支持
                checkRequest(request);
                // Execute invokeHandlerMethod in synchronized block if 
required.
                //判断是否需要在 synchronize 块中执行
    if (this.synchronizeOnSession) {
                    HttpSession session = request.getSession(false);
                    if (session != null) {
                        Object mutex = WebUtils.getSessionMutex(session);
                        synchronized (mutex) {
```

```
                    mav = invokeHandlerMethod(request, response,
handlerMethod);
                }
            }
            else {
                // No HttpSession available -> no mutex necessary
                mav = invokeHandlerMethod(request, response,
handlerMethod);
            }
        }
        else {
            // No synchronization on session demanded at all...
            mav = invokeHandlerMethod(request, response, handlerMethod);
        }
        //省略代码
        return mav;
    }

}
```

从上述代码可知，RequestMappingHandlerAdapter 的处理逻辑主要由 handleInternal()实现，而核心处理逻辑由方法 invokeHandlerMethod()实现，invokeHandlerMethod 方法的源码如下：

```
//调用处理器方法，即要执行的 Controller 中的具体的方法
protected ModelAndView invokeHandlerMethod(HttpServletRequest request,
        HttpServletResponse response, HandlerMethod handlerMethod) throws
Exception {
    ServletWebRequest webRequest = new ServletWebRequest(request,
response);
        try {
        //绑定数据
            WebDataBinderFactory binderFactory =
getDataBinderFactory(handlerMethod);
            ModelFactory modelFactory = getModelFactory(handlerMethod,
binderFactory);
    ServletInvocableHandlerMethod invocableMethod =
                        createInvocableHandlerMethod(handlerMethod);
            if (this.argumentResolvers != null) {
                invocableMethod.
    setHandlerMethodArgumentResolvers(this.argumentResolvers);
            }
            if (this.returnValueHandlers != null) {
                invocableMethod.
    setHandlerMethodReturnValueHandlers(this.returnValueHandlers);
            }
```

```
            invocableMethod.setDataBinderFactory(binderFactory);
            invocableMethod
    .setParameterNameDiscoverer(this.parameterNameDiscoverer);
         //创建 ModelAndView 容器
            ModelAndViewContainer mavContainer = new
ModelAndViewContainer();
            mavContainer.addAllAttributes
    (RequestContextUtils.getInputFlashMap(request));
            //初始化 model
    modelFactory.initModel(webRequest, mavContainer, invocableMethod);
            mavContainer.setIgnoreDefaultModelOnRedirect
    (this.ignoreDefaultModelOnRedirect);
            AsyncWebRequest asyncWebRequest =
    WebAsyncUtils.createAsyncWebRequest(request, response);
            asyncWebRequest.setTimeout(this.asyncRequestTimeout);
            WebAsyncManager asyncManager =
WebAsyncUtils.getAsyncManager(request);
            asyncManager.setTaskExecutor(this.taskExecutor);
            asyncManager.setAsyncWebRequest(asyncWebRequest);
            asyncManager.registerCallableInterceptors
(this.callableInterceptors);
            asyncManager.registerDeferredResultInterceptors
    (this.deferredResultInterceptors);
            if (asyncManager.hasConcurrentResult()) {
               Object result = asyncManager.getConcurrentResult();
               mavContainer =
    (ModelAndViewContainer) asyncManager.getConcurrentResultContext()[0];
               asyncManager.clearConcurrentResult();
               if (logger.isDebugEnabled()) {
                   logger.debug("Found concurrent result value
                   [" + result + "]");
               }
               invocableMethod = invocableMethod.
wrapConcurrentResult(result);
            }
         //执行处理器的方法
            invocableMethod.invokeAndHandle(webRequest, mavContainer);
            if (asyncManager.isConcurrentHandlingStarted()) {
               return null;
            }
         //返回 ModelAndView
            return getModelAndView(mavContainer, modelFactory, webRequest);
         }finally {
            webRequest.requestCompleted();
         }
```

```
    }
```

从上述代码可知，RequestMappingHandlerAdapter 内部对于每个请求都会实例化一个 ServletInvocableHandlerMethod（InvocableHandlerMethod 的子类）进行处理。ServletInvocableHandlerMethod 类的继承关系如图 8-5 所示。

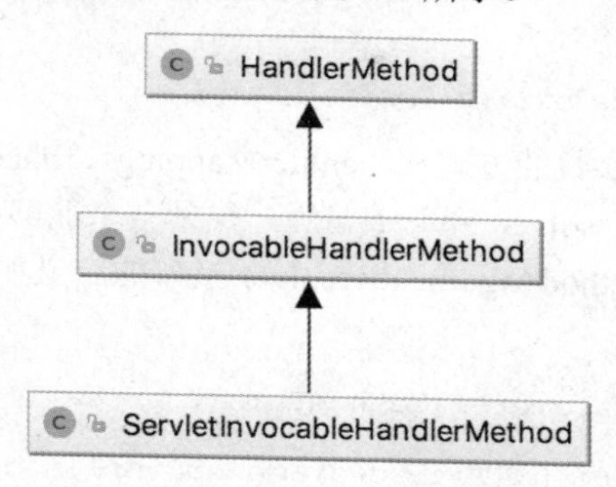

图 8-5　ServletInvocableHandlerMethod 类的继承关系

InvocableHadlerMethod 类通过调用 getMethodArgumentValues()获取方法的输入参数，具体源码如下：

```
    private Object[] getMethodArgumentValues(NativeWebRequest request,
    @Nullable ModelAndViewContainer mavContainer,
     Object... providedArgs) throws Exception {
        MethodParameter[] parameters = this.getMethodParameters();
        Object[] args = new Object[parameters.length];
        for(int i = 0; i < parameters.length; ++i) {
            MethodParameter parameter = parameters[i];
            parameter.initParameterNameDiscovery(this.parameterNameDiscoverer);
            args[i] = this.resolveProvidedArgument(parameter, providedArgs);
            if (args[i] == null) {
                //获取能够处理入参的 ArgumentResolver，然后解析参数
                if (this.argumentResolvers.supportsParameter(parameter)) {
                    try {
                        args[i] = this.argumentResolvers.resolveArgument(parameter,
mavContainer, request, this.dataBinderFactory);
                    } catch (Exception var9) {
                        if (this.logger.isDebugEnabled()) {
                            this.logger.debug
(this.getArgumentResolutionErrorMessage("Failed to resolve", i), var9);
                        }
                        throw var9;
                    }
                } else if (args[i] == null) {
                    throw new IllegalStateException("Could not resolve method
                      parameter at index " + parameter.getParameterIndex() + " in "
                      + parameter.getExecutable().toGenericString() + ": "
```

```
                +this.getArgumentResolutionErrorMessage("No suitable
                resolver for", i));
            }
        }
    }
    return args;
}
```

从上述代码可知，解析参数的方式和 handlerMappings、handlerAdapters 类似，都是从一个 HandlerMethodArgumentResolver 列表中遍历，找到一个能够处理的 bean，然后调用 bean 的核心方法处理。HandlerMethodArgumentResolver 接口的定义如下：

```
public interface HandlerMethodArgumentResolver {
    boolean supportsParameter(MethodParameter var1);

    Object resolveArgument(MethodParameter var1,ModelAndViewContainer var2,
 NativeWebRequest var3, WebDataBinderFactory var4) throws Exception;
}
```

HandlerMethodArgumentResolver 类通过 supportsParameter 筛选符合条件的 resolver，然后调用 resolver 的 resolveArgument 解析前端参数。Spring 提供了许多 HandlerMethodArgumentResolver，具体可以在 RequestMappingHandlerAdapter.afterPropertiesSet() 方法中查看。

```
private List<HandlerMethodArgumentResolver> getDefaultArgumentResolvers() {
        List<HandlerMethodArgumentResolver> resolvers = new ArrayList<>();
        // Annotation-based argument resolution
        resolvers.add(new RequestParamMethodArgumentResolver
(getBeanFactory(), false));
        resolvers.add(new RequestParamMapMethodArgumentResolver());
        resolvers.add(new PathVariableMethodArgumentResolver());
        resolvers.add(new PathVariableMapMethodArgumentResolver());
        resolvers.add(new MatrixVariableMethodArgumentResolver());
        resolvers.add(new MatrixVariableMapMethodArgumentResolver());
        resolvers.add(new ServletModelAttributeMethodProcessor(false));
        resolvers.add(new RequestResponseBodyMethodProcessor
            (getMessageConverters(), this.requestResponseBodyAdvice));
        resolvers.add(new RequestPartMethodArgumentResolver
            (getMessageConverters(), this.requestResponseBodyAdvice));
        resolvers.add(new RequestHeaderMethodArgumentResolver
            (getBeanFactory()));
        resolvers.add(new RequestHeaderMapMethodArgumentResolver());
        resolvers.add(new ServletCookieValueMethodArgumentResolver
            (getBeanFactory()));
        resolvers.add(new ExpressionValueMethodArgumentResolver
            (getBeanFactory()));
        resolvers.add(new SessionAttributeMethodArgumentResolver());
```

```
        resolvers.add(new RequestAttributeMethodArgumentResolver());

        // Type-based argument resolution
        resolvers.add(new ServletRequestMethodArgumentResolver());
        resolvers.add(new ServletResponseMethodArgumentResolver());
        resolvers.add(new HttpEntityMethodProcessor
            (getMessageConverters(), this.requestResponseBodyAdvice));
        resolvers.add(new RedirectAttributesMethodArgumentResolver());
        resolvers.add(new ModelMethodProcessor());
        resolvers.add(new MapMethodProcessor());
        resolvers.add(new ErrorsMethodArgumentResolver());
        resolvers.add(new SessionStatusMethodArgumentResolver());
        resolvers.add(new UriComponentsBuilderMethodArgumentResolver());

        // Custom arguments
        if (getCustomArgumentResolvers() != null) {
            resolvers.addAll(getCustomArgumentResolvers());
        }

        // Catch-all
        resolvers.add(new RequestParamMethodArgumentResolver
                (getBeanFactory(), true));
        resolvers.add(new ServletModelAttributeMethodProcessor(true));

        return resolvers;
    }
```

从上述代码可知，除了 Spring 提供的 RequestParamMethodArgumentResolver、PathVariableMethodArgumentResolver、SessionAttributeMethodArgumentResolver 等默认 resolver 之外，还可以自定义 resolver，通过注解来指定处理的参数类型，然后通过 getCustomArgumentResolvers 方法会注册到 revolver 列表。下面以 RequestParamMethod-ArgumentResolver 为例做简单的分析，具体类的继承关系如图 8-6 所示。

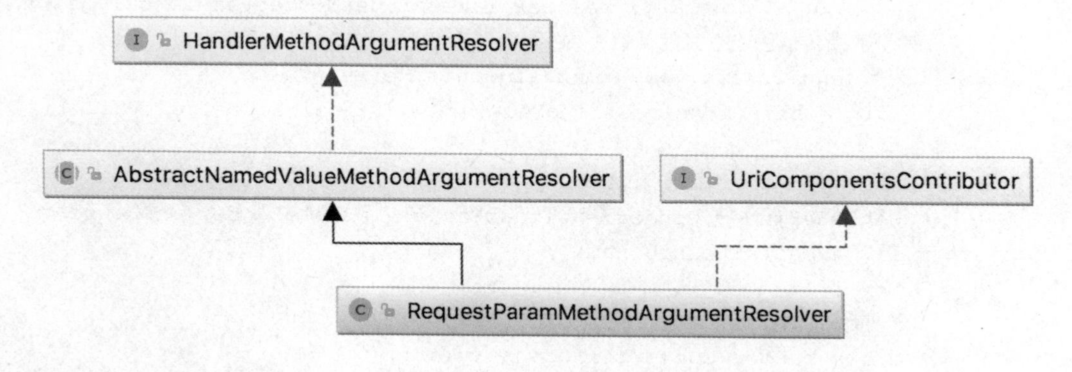

图 8-6　ServletInvocableHandlerMethod 类的继承关系

RequestParamMethodArgumentResolver 父类是 AbstractNamedValueMethodArgumentResolver，其中最核心的方法是 resolveArgument：

```
    public final Object resolveArgument(MethodParameter parameter,
    ModelAndViewContainer mavContainer, NativeWebRequest webRequest,
    WebDataBinderFactory binderFactory) throws Exception {
        AbstractNamedValueMethodArgumentResolver.NamedValueInfo namedValueInfo
    = this.getNamedValueInfo(parameter);
        MethodParameter nestedParameter = parameter.nestedIfOptional();
        //从 request 请求中解析参数的名称
    Object resolvedName = this.resolveStringValue(namedValueInfo.name);
        if (resolvedName == null) {
            throw new IllegalArgumentException("Specified name must not resolve
to null:
    [" + namedValueInfo.name + "]");
        } else {
            //通过参数名称获取参数值
            Object arg =
    this.resolveName(resolvedName.toString(), nestedParameter, webRequest);
        //判读参数值是否为空
    if (arg == null) {
            //判断默认值是否为空
            if (namedValueInfo.defaultValue != null) {
                //如果设置默认值 defaultValue，获取默认值
                arg = this.resolveStringValue(namedValueInfo.defaultValue);
            //判断是否为必填选项
            } else if (namedValueInfo.required
&& !nestedParameter.isOptional()){
                //如果是必填选项，调用 handleMissingValue，处理必填选项前端无传值情况
                this.handleMissingValue(namedValueInfo.name, nestedParameter,
webRequest);
            }
                arg = this.handleNullValue(namedValueInfo.name,
                            arg, nestedParameter.getNestedParameterType());
            //如果前端传值为“”且默认值不为空
    } else if ("".equals(arg) && namedValueInfo.defaultValue != null) {
                arg = this.resolveStringValue(namedValueInfo.defaultValue);
            }
            if (binderFactory != null) {
                WebDataBinder binder = binderFactory.createBinder(webRequest,
    (Object)null, namedValueInfo.name);
                try {
                //重点：真正进行类型转换的逻辑
                arg = binder.convertIfNecessary(arg,
    parameter.getParameterType(), parameter);
                } catch (ConversionNotSupportedException var11) {
                    //省略代码
                } catch (TypeMismatchException var12) {
```

```
                //省略代码
            }
        }
        this.handleResolvedValue(arg, namedValueInfo.name, parameter,
                                               mavContainer, webRequest);
        return arg;
    }
}
```

由上述代码可知，Spring MVC 框架将 ServletRequest 对象及处理方法的参数对象实例传递给 DataBinder，DataBinder 会调用装配在 Spring MVC 上下文的 ConversionService 组件进行数据类型和数据格式转换工作，并将 ServletRequest 中的消息填充到参数对象中。然后再调用 Validator 组件对绑定了请求消息数据的参数对象进行数据合法性校验，并最终生成数据绑定结果 BindingResult 对象。BindingResult 包含已完成数据绑定的参数对象，还包含相应的检验错误对象。

8.4 视图解析器

视图解析器（ViewResolver）是 Spring MVC 处理流程中的最后一个环节，Spring MVC 流程最后返回给用户的视图为具体的 View 对象，View 对象包含 Model 对象，而 Model 对象存放后端需要反馈给前端的数据。视图解析器把一个逻辑上的视图名称解析为一个具体的 View 视图对象，最终的视图可以是 JSP、Excel、JFreeChart 等。

8.4.1 视图解析流程

SpringMVC 的视图解析流程为：

（1）SpringMVC 调用目标方法，将目标方法返回的 String、View、ModelMap 或 ModelAndView 都转换为一个 ModelAndView 对象。

（2）通过 ViewResolver 对 ModelAndView 对象中的 View 对象进行解析，将逻辑视图 View 对象解析为一个物理视图 View 对象。

（3）调用物理视图 View 对象的 render()方法进行视图渲染，得到响应结果。

8.4.2 常用视图解析器

Spring MVC 提供了很多视图解析器类，具体如图 8-7 所示。

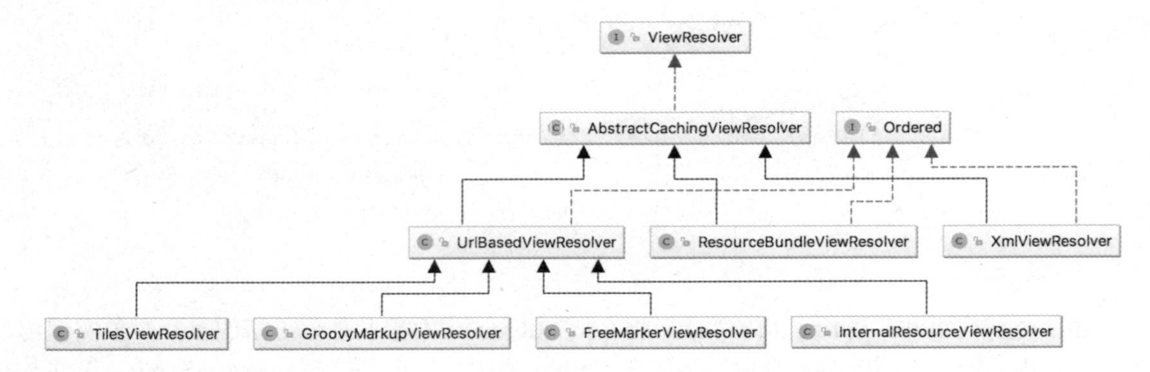

图 8-7　ViewResolver 类继承关系

下面介绍一些常用的视图解析器类。

1. ViewResolver

ViewResolver 是所有视图解析器的父类，具体源码如下：

```
public interface ViewResolver {
    @Nullable
    View resolveViewName(String viewName, Locale locale) throws Exception;
}
```

ViewResolver 的主要作用是把一个逻辑上的视图名称解析为一个真正的视图，然后通过 View 对象进行渲染。

2. AbstractCachingViewResolver

抽象类，这种视图解析器会把解析过的视图保存起来，然后在每次解析视图时先从缓存里面查找，如果找到了对应的视图就直接返回，如果没有找到就创建一个新的视图对象，然后把它存放到用于缓存的 Map 中，接着再把新建的视图返回。使用这种视图缓存的方式可以把解析视图的性能问题降到最低。

3. UrlBasedViewResolver

该类继承了 AbstractCachingViewResolver，主要是提供一种拼接 URL 的方式来解析视图，它可以让我们将 prefix 属性指定的前缀通过 suffix 属性指定后缀，然后把返回的逻辑视图名称加上指定的前缀和后缀就是指定的视图 URL 了。如 prefix=/WEB-INF/jsps/，suffix=.jsp，返回的视图名称为 viewName=test/indx，则 UrlBasedViewResolver 解析出来的视图 URL 就是/WEB-INF/jsps/test/ index.jsp，默认的 prefix 和 suffix 都是空串。

URLBasedViewResolver 支持返回的视图名称中包含 redirect:前缀，这样就可以支持 URL 在客户端的跳转，如当返回的视图名称是“redirect:test.do”的时候，URLBasedViewResolver 发现返回的视图名称包含“redirect:”前缀，于是把返回的视图名称前缀“redirect:”去掉，取后面的 test.do 组成一个 RedirectView，RedirectView 中将把请求返回的模型属性组合成查询参数的形式再组合到 redirect 的 URL 后面，然后调用 HttpServletResponse 对象的 sendRedirect 方法进行重定向。同样 URLBasedViewResolver 还支持 forword:前缀，对于视图名称中包含 forword:前缀的视图名称将会被封装成一个 InternalResourceView 对象，然后在服务器端利用 RequestDispatcher 的 forword 方式跳转到指定的地址。使用 UrlBasedViewResolver 的时候必须指定属性 viewClass，表示解析成哪种视图，一般使用较多的就是 InternalResourceView，利用它来展现 JSP，但是当使用 JSTL 的时候必须使用 JstlView。具体示例如下：

```
<bean
    class="org.springframework.web.servlet.view.UrlBasedViewResolver">
    <property name="prefix" value="/WEB-INF/" />
    <property name="suffix" value=".jsp" />
    <property name="viewClass"
value="org.springframework.web.servlet.view.InternalResourceView"/>
</bean>
```

上述代码中，当返回的逻辑视图名称为 test 时，UrlBasedViewResolver 将逻辑视图名称加上定义好的前缀和后缀，即“/WEB-INF/test.jsp”，然后新建一个 viewClass 属性指定的视图类型予以返回，即返回一个 URL 为“/WEB-INF/test.jsp”的 InternalResourceView 对象。

4. InternalResourceViewResolver

该类是 URLBasedViewResolver 的子类，所以 URLBasedViewResolver 支持的特性它都支持。InternalResourceViewResolver 是使用最广泛的一个视图解析器，可以把 InternalResourceViewResolver 解释为内部资源视图解析器，InternalResourceViewResolver 会把返回的视图名称都解析为 InternalResourceView 对象，InternalResourceView 会把 Controller 处理器方法返回的模型属性都存放到对应的 request 属性中，然后通过 RequestDispatcher 在服务器端把请求 forword 重定向到目标 URL。比如在 InternalResourceViewResolver 中定义了 prefix=/WEB-INF/，suffix=.jsp，然后请求的 Controller 处理器方法返回的视图名称为 test，那么这个时候 InternalResourceViewResolver 就会把 test 解析为一个 InternalResourceView 对象，先把返回的模型属性都存放到对应的 HttpServletRequest 属性中，然后利用 RequestDispatcher 在服务器端把请求 forword 到/WEB-INF/test.jsp。这就是 InternalResourceViewResolver 一个非常重要的特性。

我们知道，存放在/WEB-INF/下面的内容是不能直接通过 request 请求的方式请求到的，为了安全性考虑，通常会把 JSP 文件放在 WEB-INF 目录下，而 InternalResourceView 在服务器端跳转的方式可以很好地解决这个问题。

```
<bean class="org.springframework.web.serolet.view.
InternalResourceViewResolver">
    <property name="prefix" value="/WEB-INF/"/>
    <property name="suffix" value=".jsp"></property>
</bean>
```

上述代码是一个 InternalResourceViewResolver 的定义，根据该定义当返回的逻辑视图名称是 test 的时候，InternalResourceViewResolver 会给它加上定义好的前缀和后缀，组成“/WEB-INF/test.jsp”的形式，然后把它当作一个 InternalResourceView 的 URL 新建一个

InternalResourceView 对象返回。

5. XmlViewResolver

它继承自 AbstractCachingViewResolver 抽象类，所以它也是支持视图缓存的。XmlViewResolver 需要给定一个 XML 配置文件，该文件将使用和 Spring 的 Bean 工厂配置文件一样的 DTD 定义，所以其实该文件就是用来定义视图的 Bean 对象的。在该文件中定义的每一个视图的 Bean 对象都给定一个名字，然后 XmlViewResolver 将根据 Controller 处理器方法返回的逻辑视图名称到 XmlViewResolver 指定的配置文件中寻找对应名称的视图 Bean 用于处理视图。该配置文件默认是/WEB-INF/views.xml 文件，如果不使用默认值可以在 XmlViewResolver 的 location 属性中指定它的位置。XmlViewResolver 还实现了 Ordered 接口，因此可以通过其 order 属性来指定在 ViewResolver 链中它所处的位置，order 的值越小优先级越高。以下是使用 XmlViewResolver 的一个示例：

```
<bean class="org.springframework.web.servlet.view.XmlViewResolver">
   <property name="location" value="/WEB-INF/views.xml"/>
   <property name="order" value="1"/>
</bean>
```

在 Spring MVC 的配置文件中加入 XmlViewResolver 的 bean 定义。使用 location 属性指定其配置文件所在的位置，order 属性指定当有多个 ViewResolver 的时候其处理视图的优先级。

在 XmlViewResolver 对应的配置文件中配置好所需要的视图定义，视图配置文件 views.xml 具体的配置如下所示：

```
<?xml version="1.0" encoding="UTF-8"?>
   <beans xmlns="http://www.springframework.org/schema/beans"
   xmlns:xsi="http://www.w3.org/2001/XMLSchema-instance"
   xsi:schemaLocation="http://www.springframework.org/schema/beans
   http://www.springframework.org/schema/beans/spring-beans-3.0.xsd">
   <bean id="index"
 class="org.springframework.web.servlet.view.InternalResourceView">
      <property name="url" value="/index.jsp"/>
   </bean>
</beans>
```

最后，定义一个返回的逻辑视图名称为在 XmlViewResolver 配置文件中定义的视图名称 index：

```
@RequestMapping("/index")
public String index() {
   return "index";
}
```

当访问上面定义好的 index 方法的时候返回的逻辑视图名称为“index”，这时候 Spring MVC 会从 views.xml 配置文件中寻找 id 或者 name 为“index”的 Bean 对象予以返回，这里

Spring 找到的是一个 URL 为“/index.jsp”的 InternalResourceView 对象，然后进行视图解析，将最终的视图页面显示给用户。

6. BeanNameViewResolver

这个视图解析器与 XmlViewResolver 有点类似，也是通过把返回的逻辑视图名称匹配定义好的视图 Bean 对象。主要的区别有两点：

（1）BeanNameViewResolver 要求视图 Bean 对象都定义在 Spring 的 application context 中，而 XmlViewResolver 是在指定的配置文件中寻找视图 Bean 对象。

（2）BeanNameViewResolver 不会进行视图缓存。

下面来看一个具体的示例：

```
<bean class="org.springframework.web.servlet.view.BeanNameViewResolver">
    <property name="order" value="1"/>
</bean>
<bean id="test" class="org.springframework.web.servlet.view.
InternalResourceView">
    <property name="url" value="/index.jsp"/>
</bean>
```

上述代码中，在 Spring MVC 的配置文件中定义了一个 BeanNameViewResolver 视图解析器和一个 id 为 test 的 InternalResourceview bean 对象。这样当返回的逻辑视图名称为 test 时，就会解析为上面定义好的 id 为 test 的 InternalResourceView 对象，然后跳转到 index.jsp 页面。

7. ResourceBundleViewResolver

该类也是继承自 AbstractCachingViewResolver 类，但是它缓存的不是视图。和 XmlViewResolver 一样，它也需要有一个配置文件来定义逻辑视图名称和真正的 View 对象的对应关系，不同的是 ResourceBundleViewResolver 的配置文件是一个属性文件，而且必须是放在 classpath 路径下面的，默认情况下这个配置文件是在 classpath 根目录下的 views.properties 文件，如果不使用默认值，则可以通过属性 baseName 或 baseNames 来指定。baseName 只是指定一个基名称，Spring 会在指定的 classpath 根目录下寻找已指定的 baseName 开始的属性文件进行 View 解析，如指定的 baseName 是 base，那么 base.properties、baseabc.properties 等以 base 开始的属性文件都会被 Spring 当作 ResourceBundleViewResolver 解析视图的资源文件。ResourceBundleViewResolver 使用的属性配置文件的内容类似于这样：

```
resourceBundle.(class)=org.springframework.web.servlet.view.InternalResour
ceView
    resourceBundle.url=/index.jsp
    test.(class)=org.springframework.web.servlet.view.InternalResourceView
    test.url=/test.jsp
```

在这个配置文件中定义了两个 InternalResourceView 对象，一个名称是 resourceBundle，对应的 URL 是/index.jsp，另一个名称是 test，对应的 URL 是/test.jsp。从这个定义可以知道，resourceBundle 对应的是视图名称，使用 resourceBundle.(class)来指定它对应的视图类型，resourceBundle.url 指定这个视图的 URL 属性。

读者可以看到，resourceBundle 的 class 属性要用小括号包起来，而它的 URL 属性为什么不需要呢？这就需要从 ResourceBundleViewResolver 进行视图解析的方法来说明。ResourceBundleViewResolver 还是通过 Bean 工厂来获得对应视图名称的视图 Bean 对象来解析视图的，那么这些 Bean 从哪里来呢？就是从我们定义的 properties 属性文件中来。在 ResourceBundleViewResolver 第一次进行视图解析的时候会先 new 一个 BeanFactory 对象，然后把 properties 文件中定义好的属性按照它自身的规则生成一个个的 Bean 对象注册到该 BeanFactory 中，之后会把该 BeanFactory 对象保存起来，所以 ResourceBundleViewResolver 缓存的是 BeanFactory，而不是直接缓存从 BeanFactory 中取出的视图 Bean。然后会从 Bean 工厂中取出名称为逻辑视图名称的视图 Bean 进行返回。

接下来介绍 Spring 通过 properties 文件生成 Bean 的规则。它会把 properties 文件中定义的属性名称按最后一个点“.”进行分割，把点前面的内容当作是 Bean 名称，点后面的内容当作是 Bean 的属性。这其中有几个特别的属性，Spring 把它们用小括号包起来了，这些特殊的属性一般是对应的 attribute，但不是 Bean 对象所有的 attribute 都可以这样用。其中（class）是一个，除了（class）之外，还有（scope）、（parent）、（abstract）、（lazy-init）。而除了这些特殊的属性之外的其他属性，Spring 会把它们当作 Bean 对象的一般属性进行处理，就是 Bean 对象对应的 property。所以根据上面的属性配置文件将生成如下两个 Bean 对象：

```
<bean id="resourceBundle"
class="org.springframework.web.servlet.view.InternalResourceView">
   <property name="url" value="/index.jsp"/>
</bean>
<bean id="test"
class="org.springframework.web.servlet.view.InternalResourceView">
   <property name="url" value="/test.jsp"/>
</bean>
```

8. FreeMarkerViewResolver

FreeMarkerViewResolver 是 UrlBasedViewResolver 的一个子类，它会把 Controller 处理方法返回的逻辑视图解析为 FreeMarkerView。FreeMarkerViewResolver 会按照 UrlBasedViewResolver 拼接 URL 的方式进行视图路径的解析，但是使用 FreeMarkerViewResolver 的时候不需要指定其 viewClass，因为 FreeMarkerViewResolver 中已经把 viewClass 定死为 FreeMarkerView 了。我们先在 Spring MVC 的配置文件里面定义一个 FreeMarkerViewResolver 视图解析器，并定义其解析视图的 order 顺序为 1，代码示例如下：

```
<bean
class="org.springframework.web.servlet.view.freemarker.FreeMarkerViewResol
ver">
```

```
        <property name="prefix" value="fm_"/>
        <property name="suffix" value=".ftl"/>
        <property name="order" value="1"/>
    </bean>
```

当请求的处理器方法返回一个逻辑视图名称 viewName 的时候，就会被该视图处理器加上前后缀解析为一个 URL 为“fm_viewName.ftl”的 FreeMarkerView 对象。对于 FreeMarkerView 需要给定一个 FreeMarkerConfig 的 Bean 对象来定义 FreeMarker 的配置信息。FreeMarkerConfig 是一个接口，Spring 已经为我们提供了一个实现，它就是 FreeMarkerConfigurer。可以通过在 Spring MVC 的配置文件里定义该 Bean 对象来定义 FreeMarker 的配置信息，该配置信息将会在 FreeMarkerView 进行渲染的时候使用到。对于 FreeMarkerConfigurer 而言，最简单的就是配置一个 templateLoaderPath，告诉 Spring 应该到哪里寻找 FreeMarker 的模板文件。这个 templateLoaderPath 也支持使用“classpath:”和“file:”前缀。当 FreeMarker 的模板文件放在多个不同的路径下面的时候，可以使用 templateLoaderPaths 属性来指定多个路径。在这里指定模板文件放在“/WEB-INF/freemarker/template”下面，示例代码如下：

```
    <bean
    class="org.springframework.web.servlet.view.freemarker.FreeMarkerConfigurer">
        <property name="templateLoaderPath" value="/WEB-INF/freemarker/
template"/>
    </bean>
```

8.4.3 ViewResolver 链

在 Spring MVC 中可以同时定义多个 ViewResolver 视图解析器，然后它们会组成一个 ViewResolver 链。当 Controller 处理器方法返回一个逻辑视图名称后，ViewResolver 链将根据其中 ViewResolver 的优先级来进行处理。所有的 ViewResolver 都实现了 Ordered 接口，在 Spring 中实现了这个接口的类都是可以排序的。ViewResolver 是通过 order 属性来指定顺序的，默认都是最大值。所以可以通过指定 ViewResolver 的 order 属性来实现 ViewResolver 的优先级，order 属性是 Integer 类型，order 越小优先级越高，所以第一个进行解析的将是 ViewResolver 链中 order 值最小的那个。

如果 ViewResolver 进行视图解析后返回的 View 对象为 null，则表示 ViewResolver 不能解析该视图，这个时候如果还存在其他 order 值比它大的 ViewResolver，就会调用剩余的 ViewResolver 中 order 值最小的那个来解析该视图，依此类推。当 ViewResolver 在进行视图解析后返回的是一个非空的 View 对象的时候，则表示该 ViewResolver 能够解析该视图，那么视图解析就完成了，后续的 ViewResolver 将不会再用来解析该视图。当定义的所有 ViewResolver 都不能解析该视图的时候，Spring 就会抛出一个异常。

基于 Spring 支持的这种 ViewResolver 链模式，就可以在 Spring MVC 应用中同时定义多个 ViewResolver，给定不同的 order 值，这样就可以对特定的视图进行处理，以此来支持同一应用中有多种视图类型。

像 InternalResourceViewResolver 这种能解析所有的视图，即永远能返回一个非空 View 对象的 ViewResolver，一定要把它放在 ViewResolver 链的最后面。

8.5 思考与练习

1. 简述 Spring MVC 的执行流程。
2. 简述前端控制器 DispatcherSerrvlet 的作用。
3. 简述 DispatcherSerrvlet 类的继承结构。
4. Spring MVC 处理映射器的含义是什么？
5. Spring MVC 提供了哪些处理映射器？
6. 简述 SimpleUrlHandlerMapping 处理映射器的作用。
7. Spring MVC 提供了哪些处理适配器？
8. 简述 Spring MVC 视图解析流程。
9. 简述 Spring MVC 常用的视图解析器。

第9章 MyBatis原理剖析

本章主要介绍 MyBatis 的整体框架、MyBatis 的初始化流程和原理以及 MyBatis 的执行流程和原理等。

9.1 MyBatis 的整体框架介绍

MyBatis 整体架构分为三层，分别是基础支持层、核心处理层和接口层，具体如图 9-1 所示。

9.1.1 接口层

接口层是上层运用与 MyBatis 交互的桥梁，其核心是 SqlSession 接口。SqlSession 接口暴露了一系列的增删改查等 API 给应用程序。接口层在接收到调用请求时，会调用核心处理层的相应模块完成具体的数据库操作。MySQL 提供了两个 SqlSession 接口的实现，具体如图 9-2 所示。

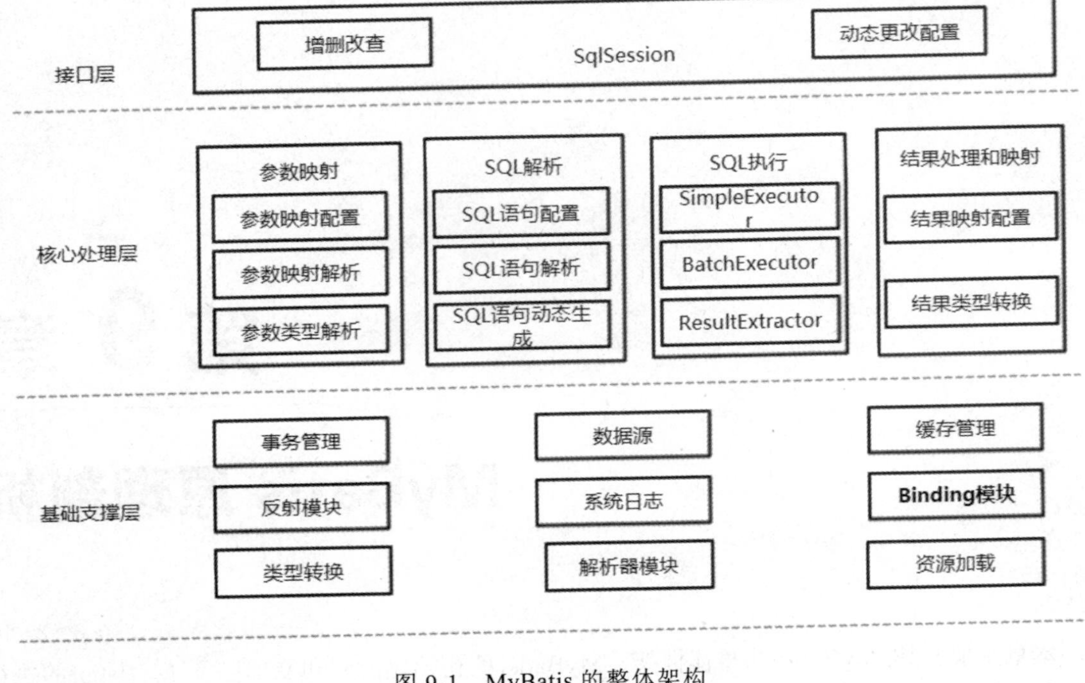

图 9-1　MyBatis 的整体架构

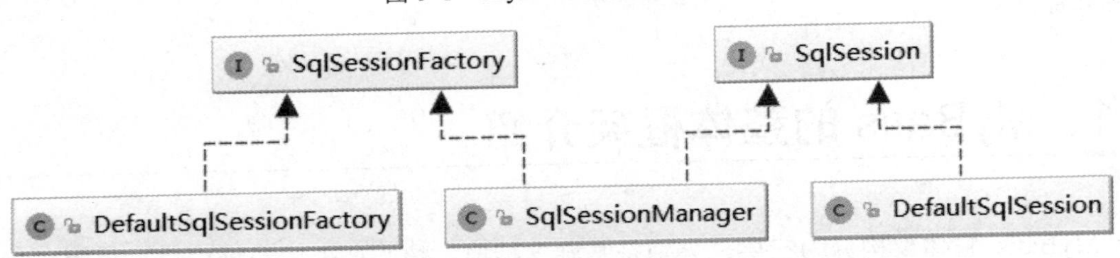

图 9-2　SqlSession 接口实现

由图 9-2 可知，SqlSession 接口实现使用了工厂方法模式，SqlSessionFactory 负责创建 SqlSession 对象，其包含多个 openSession()方法的重载，可以通过参数指定事务的隔离级别 TransactionIsolationLevel、底层使用 Excutor 的类型以及是否自动提交事务等方面的配置。SqlSessionFactory 的源码如下：

```
public interface SqlSessionFactory {

  SqlSession openSession();

  SqlSession openSession(boolean autoCommit);
  SqlSession openSession(Connection connection);
  SqlSession openSession(TransactionIsolationLevel level);

  SqlSession openSession(ExecutorType execType);
  SqlSession openSession(ExecutorType execType, boolean autoCommit);
  SqlSession openSession(ExecutorType execType, TransactionIsolationLevel
```

```
level);
      SqlSession openSession(ExecutorType execType, Connection connection);

      Configuration getConfiguration();

   }
```

SqlSession 接口提供了常用的增删改查的数据库操作以及事务的相关操作。同时，每种类型的操作都提供了多种重载。SqlSessionFactory 的源码如下：

```
   public interface SqlSession extends Closeable {
      //查询方法：使用 SQL 语句查询，返回值为查询的结果对象
      <T> T selectOne(String statement);
      //查询方法：使用 SQL 语句查询，第二个参数表示用户传入的参数，也就是 SQL 语句绑定的实参
      <T> T selectOne(String statement, Object parameter);
      //查询方法：用来查询多条记录，查询结果封装成结果对象列表返回
      <E> List<E> selectList(String statement);
      <E> List<E> selectList(String statement, Object parameter);
      //带分页的查询方法：第三个参数用来限制解析结果集的范围
      <E> List<E> selectList(String statement, Object parameter, RowBounds
rowBounds);
      //查询方法：查询结果会被映射成 Map 对象返回。其中，第二个参数指定了结果集哪一列作为 Map
的 key
      <K, V> Map<K, V> selectMap(String statement, String mapKey);
      <K, V> Map<K, V> selectMap(String statement, Object parameter, String
mapKey);
      //返回值为游标对象，参数含义与 selectList()方法相同
      <T> Cursor<T> selectCursor(String statement);
      <T> Cursor<T> selectCursor(String statement, Object parameter);
      <T> Cursor<T> selectCursor(String statement, Object parameter, RowBounds
rowBounds);
      //查询的结果对象将由 ResultHandler 对象处理，其余参数与 selectList()方法相同
      void select(String statement, Object parameter, ResultHandler handler);
      void select(String statement, ResultHandler handler);
      void select(String statement, Object parameter, RowBounds rowBounds,
   ResultHandler handler);
      //执行插入语句
      int insert(String statement);
      int insert(String statement, Object parameter);
      //执行更新语句
      int update(String statement);
      int update(String statement, Object parameter);
      //执行删除语句
      int delete(String statement);
      int delete(String statement, Object parameter);
      //提交事务
```

```
    void commit();
    void commit(boolean force);
    //回滚事务
    void rollback();
    void rollback(boolean force);
    //将请求刷新到缓存
    List<BatchResult> flushStatements();
    //关闭 session
    void close();
    //清空缓存
    void clearCache();
    //获取 Configuration 对象
    Configuration getConfiguration();
    //获取 type 对应的 Mapper 对象
    <T> T getMapper(Class<T> type);
     //获取 Sqlsession 对应的数据库连接
    Connection getConnection();
}
```

9.1.2 核心处理层

MyBatis 的核心处理层完成了 MyBatis 核心处理流程，包括 MyBatis 的初始化及完成一次数据库操作所涉及的全部流程。具体说明如下。

1. 参数映射

MyBatis 在初始化过程中，会加载配置文件 mybatis-config.xml、映射配置文件以及 Mapper 接口中的注解信息，解析配置文件后会生成相关的对象保存到 Configuration 对象中。例如以下代码中的<resultMap>节点会被解析成 ResultMap 对象，<result>节点会被解析成 ResultMapping 对象。利用 Configuration 对象可以创建 SqlSessionFactory 对象，并通过 SqlSessionFactory 工程对象创建 SqlSession 对象完成数据库操作。

```
     <resultMap id="userMap" type="com.ay.model.AyUser">
        <id property="id" column="id"/>
        <result property="name" column="name"/>
    </resultMap>
```

2. SQL解析

MyBatis 实现动态 SQL 语句的功能，提供了多种动态 SQL 语句对应的节点，比如<where>节点、<if>节点、<foreach>节点等。通过这些节点的组合使用，开发人员可以写出满足所有需求的动态 SQL 语句。MyBatis 会根据用户传入的实参，解析映射文件中定义的动态 SQL 节点，生成可执行的 SQL 语句。之后会处理 SQL 语句中的占位符，绑定用户传入的实参。

3. SQL执行与结果处理和映射

SQL 语句执行涉及 Executor、StatementHandler、PreparedHandler 和 ResultSetHandler。

其中 Executor 主要负责维护一级缓存和二级缓存，并提供事务管理的相关操作，它会将数据库相关操作委托给 StatementHandler 完成。StatementHandler 首先通过 PreparedHandler 完成 SQL 语句的实参绑定，然后通过 Statement 对象执行 SQL 语句并得到结果集，最后通过 ResultSetHandler 完成结果集的映射，得到结果对象并返回。

9.1.3　基础支撑层

包括以下几方面的任务：

1. 事务管理

项目开发过程中，一般很少直接使用 MyBatis 事务管理，而是 MyBatis 和 Spring 框架集成时，使用 Spring 框架管理事务。MyBatis 框架又对数据库中的事务进行了抽象，其自身提供了相应的事务接口和简单实现。

2. 数据源

顾名思义，数据源是提供某种所需要数据的器件或原始媒体。在数据源中存储了所有建立数据库连接的信息，就像通过指定文件名称可以在文件系统中找到文件一样，通过提供正确的数据源名称，你可以找到相应的数据库连接。MyBatis 提供了相应的数据源实现，当然 MyBatis 也提供了与第三方数据源集成的接口，这些功能都位于数据源模块中。

3. 缓存管理

MyBatis 中提供了一级缓存和二级缓存。需要注意的是，MyBatis 中自带的这两级缓存与 MyBatis 以及整个应用是运行在同一个 JVM 中的，共享同一块堆内存。如果这两级缓存中的数据量较大，则可能影响系统中其他功能的运行，所以当需要缓存大量数据时，建议优先考虑使用 Redis、Memcache 等缓存产品。

4. 反射模块

MyBatis 提供反射模块对 Java 原生的反射进行良好的封装，其提供了简洁的 API，方便上层使用，并且对反射操作进行一系列优化，例如缓存了类的元数据，提高了反射操作的性能。

5. 系统日志

MyBatis 日志模块的主要功能是集成第三方日志框架，例如 Log4j、Log4j2、slf4j 等。除此之外，MyBatis 还提供了详细的日志输出信息。

6. Binding模块

MyBatis 通过 Binding 模块将用户自定义的 Mapper 接口与映射配置文件关联起来，系统可以通过自定义 Mapper 接口中的方法执行相应的 SQL 语句完成数据库操作。

7. 类型转换

MyBatis 为简化配置文件提供了别名机制，该机制是类型转换模块的主要功能之一。类型转换模块的另一个功能是实现 JDBC 类型与 Java 类型之间的转换，该功能在为 SQL 语句绑定实参以及映射查询结果集时都会涉及。在为 SQL 语句绑定实参时，会将数据由 Java 类型转换为 JDBC 类型；而在映射结果集时，会将数据由 JDBC 类型转换为 Java 类型。

8. 解析器模块

MyBatis 的解析器模块主要提供了以下两个功能：

（1）对 Xpath 进行封装，为 MyBatis 初始化时解析 mybatis-config.xml 配置文件以及映射配置文件提供支持。

（2）为处理动态 SQL 语句中的占位符提供支持。

9. 资源加载

MyBatis 的资源加载模块主要是对类加载器进行封装，确定类加载器的使用顺序，并提供加载类文件以及其他资源文件的功能。

9.2 MyBatis 初始化流程

任何框架的启动无非是加载自己运行时所需要的配置信息，MyBatis 的配置信息主要在 mybatis-config.xml 文件里配置，具体代码如下：

```
    <?xml version="1.0" encoding="UTF-8" ?>
<!DOCTYPE configuration
       PUBLIC "-//mybatis.org//DTD Config 3.0//EN"
       "http://mybatis.org/dtd/mybatis-3-config.dtd">
<configuration>
    <!-- 设置 -->
<settings></settings>
<!-- 属性 -->
<properties></properties>
<!-- 类型命名 -->
<typeAliases></typeAliases>
<!-- 类型处理器 -->
<typeHandlers></typeHandlers>
<!-- 对象工厂 -->
<objectFactory type="" ></objectFactory>
<!-- 插件 -->
   <plugins>
       <plugin interceptor=""></plugin>
</plugins>
<!-- 环境 -->
<environments default="">
      <!-- 环境变量 -->
```

```
        <environment id="">
            <!-- 事务管理器 -->
            <transactionManager type=""></transactionManager>
            <!-- 数据源 -->
            <dataSource type=""></dataSource>
        </environment>
    </environments>
    <!-- 数据库厂商标识 -->
    <databaseIdProvider type=""></databaseIdProvider>
    <!-- 映射器 -->
       <mappers></mappers>
    </configuration>
```

上述的 XML 配置信息，MyBatis 使用 Configuration 对象作为配置信息的容器，Configuration 对象的组织结构和 XML 配置文件的组织结构几乎完全一样。MyBatis 完成 Configuration 对象初始化之后，用户就可以使用 MyBatis 进行数据库操作了。换言之，MyBatis 初始化的过程，就是创建 Configuration 对象的过程。

首先，来看下面的一个简单示例：

```
    String resource = "mybatis-config.xml";
    InputStream inputStream = Resources.getResourceAsStream(resource);
    //初始化流程主要代码
    SqlSessionFactory sqlSessionFactory = new
SqlSessionFactoryBuilder().build(inputStream);
    SqlSession sqlSession = sqlSessionFactory.openSession();
    List list = sqlSession.selectList("");
```

上述代码的执行过程如下：

（1）将配置文件 mybatis-config.xml 加载成 InputStream 输入流。
（2）通过 SqlSessionFactoryBuilder 的 build()方法创建 SqlSessionFactory 对象。
（3）通过 SqlSessionFactory 的 openSession()方法创建 SqlSession 对象。
（4）执行 Mapper 文件中的查询语句，获取查询结果。

MyBatis 的初始化流程主要发生在第（2）步，SqlSessionFactoryBuilder 根据传入的数据流生成 Configuration 对象，然后根据 Configuration 对象创建默认的 SqlSessionFactory 实例。初始化的基本过程如图 9-3 所示。

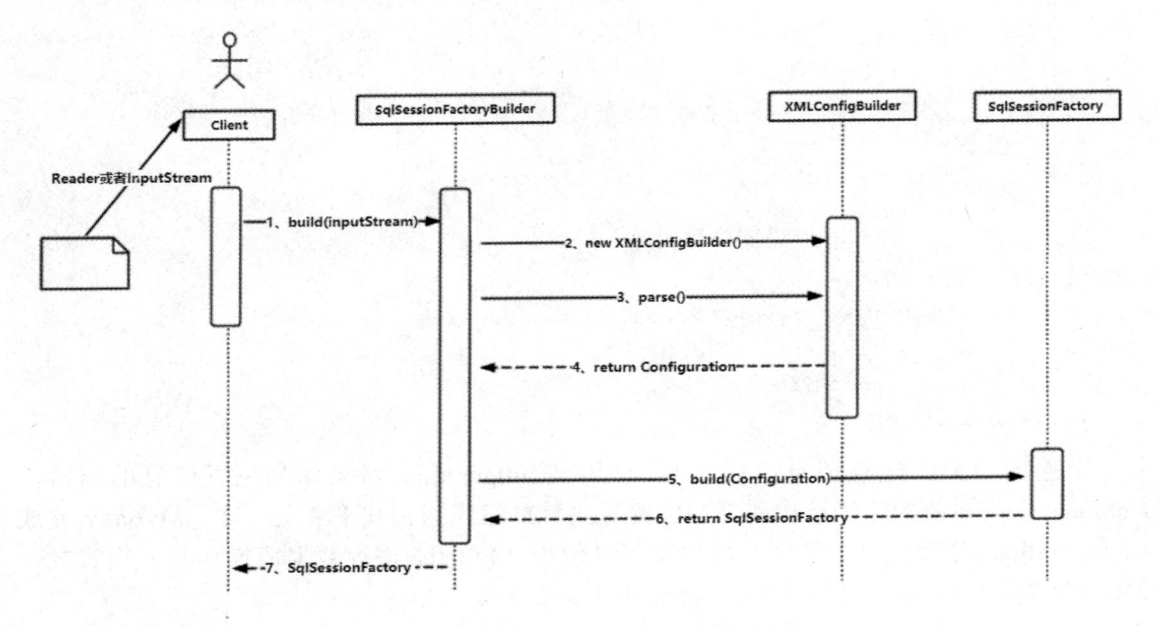

图 9-3　MyBatis 的初始化流程

由图 9-3 可知，MyBatis 的初始化要经过以下几步：

（1）客户端 Client 调用 SqlSessionFactoryBuilder 对象的 build（inputStream）方法。

（2）SqlSessionFactoryBuilder 会根据输入流 inputStream 等信息创建 XMLConfigBuilder 对象。

（3）SqlSessionFactoryBuilder 调用 XMLConfigBuilder 对象的 parse()方法。

（4）XMLConfigBuilder 对象返回 Configuration 对象。

（5）SqlSessionFactoryBuilder 根据 Configuration 对象创建一个 DefaultSessionFactory 对象。

（6）SqlSessionFactoryBuilder 返回 DefaultSessionFactory 对象给 Client，供 Client 使用。

9.3 MyBatis 的执行流程

首先，我们再来看之前这个简单示例：

```
String resource = "mybatis-config.xml";
InputStream inputStream = Resources.getResourceAsStream(resource);
//初始化流程
SqlSessionFactory sqlSessionFactory = new
SqlSessionFactoryBuilder().build(inputStream);
//执行流程
SqlSession sqlSession = sqlSessionFactory.openSession();
```

```
//执行流程
List list = sqlSession.selectList("com.ay.dao.AyUserDao. findById", param);
sqlSession.close();
```

SqlSessionFactory sqlSessionFactory = new SqlSessionFactoryBuilder().build(inputStream);主要涉及 MyBatis 的初始化流程，而 sqlSessionFactory.openSession()、sqlSession.selectList("")和 sqlSession.close()涉及 MyBatis 执行流程。MyBatis 执行流程具体如图 9-4 所示（参考网络图片）。

（1）SqlSessionFactory 调用 openSession()方法获取 SqlSession 对象，SqlSession 对象封装了对数据库的 CRUD（增删改查）和事务控制。

（2）为 SqlSession 传递一个配置 SQL 语句的 Statement Id 和参数，然后返回结果。具体代码如下：

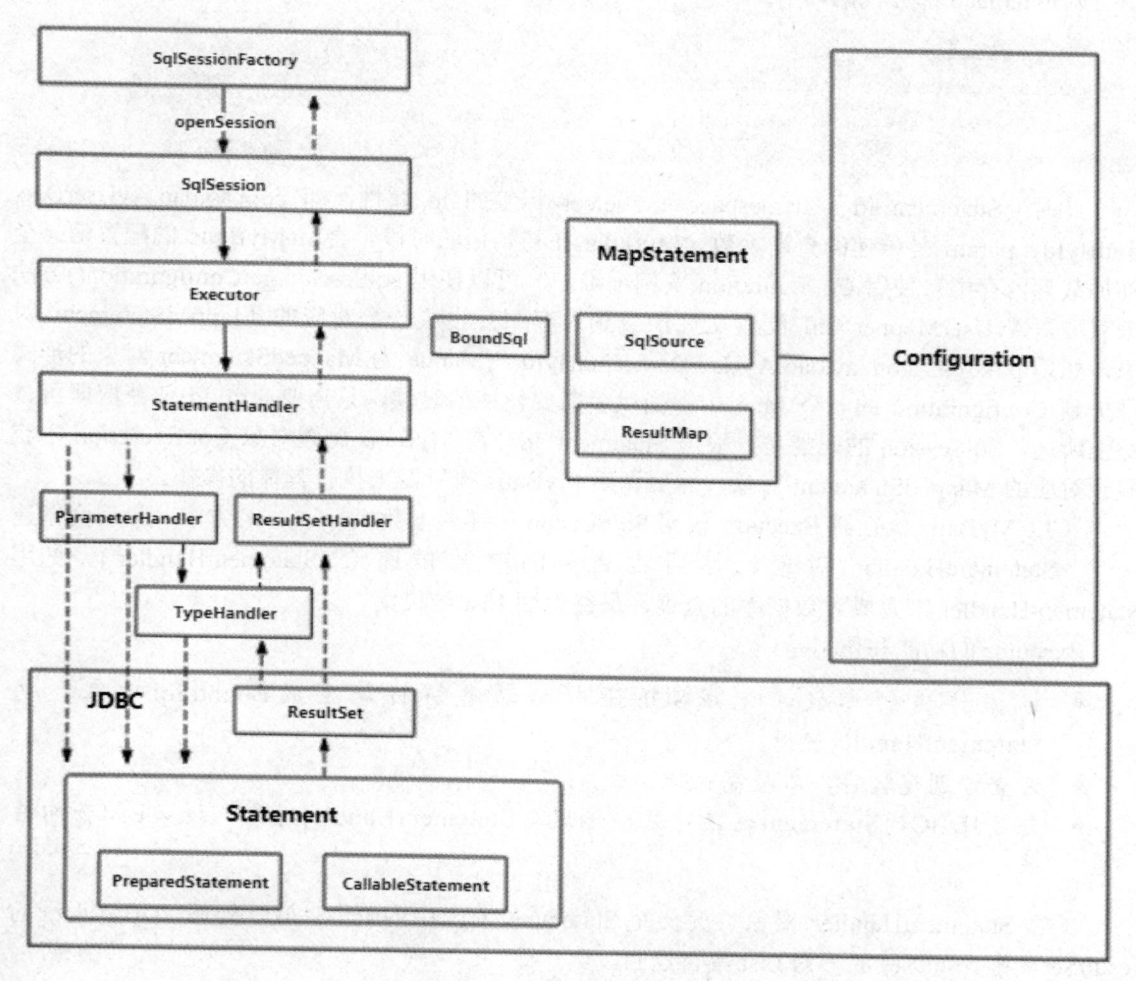

图 9-4 MyBatis 的执行流程

```
//执行流程
List list = sqlSession.selectList("com.ay.dao.AyUserDao. findById", param);
```

AyUserMapper.xml 部分代码如下：

```
<?xml version="1.0" encoding="UTF-8" ?>
<!DOCTYPE mapper PUBLIC "-//mybatis.org//DTD Mapper 3.0//EN"
    "http://mybatis.org/dtd/mybatis-3-mapper.dtd">
<mapper namespace="com.ay.dao.AyUserDao">
<select id="findById" useCache="false"  parameterType="String"
resultMap="userMap">
    SELECT * FROM ay_user
    WHERE id = #{id}
</select>
</mapper>
```

对应的 Dao 部分代码如下：

```
@Repository
public interface AyUserDao {
   AyUser findById(String id);
}
```

此时，Statement Id = namespace + <select>标签的 id 属性，即 com.ay.dao.AyUserDao.findById，param 是传递的查询参数。MyBatis 在初始化的时候，会将 MyBatis 的配置信息全部加载到内存中，使用 Configuration 实例来维护。可以使用 sqlSession.getConfiguration()方法来获取。AyUserMapper.xml 配置文件加载到内存中会生成一个对应的 MappedStatement 对象，然后以 key=" com.ay.dao.AyUserDao. findById "，value 为 MappedStatement 对象的形式维护到 Configuration 的一个 Map 中。当以后需要使用的时候，只需要通过 Id 值获取即可。综上所述，SqlSession 的职能是，根据 Statement Id，在 MyBatis 配置对象 Configuration 中获取到对应的 MappedStatement 对象，然后调用 MyBatis 执行器来执行具体的操作。

（3）MyBatis 执行器 Executor 根据 SqlSession 传递的参数执行 query()方法，最后会创建一个 StatementHandler 对象，然后将必要的参数传递给 StatementHandler，使用 StatementHandler 来完成对数据库的查询，最终返回 List 结果集。

Executor 的功能和作用是：

- 根据传递的参数，完成SQL语句的动态解析，生成BoundSql对象，供StatementHandler使用。
- 为查询创建缓存，以提高性能。
- 创建JDBC的Statement连接对象，传递给StatementHandler对象，返回List查询结果。

（4）StatementHandler 对象负责设置 Statement 对象中的查询参数、处理 JDBC 返回的 resultSet，将 resultSet 加工为 List 集合返回。

StatementHandler 对象主要完成以下两个工作：

- 对于JDBC的PreparedStatement类型的对象，在创建的过程中，使用的是SQL语句，字符串会包含若干个“?”占位符，其后再对占位符进行设值，

StatementHandler通过parameterize(statement)方法对Statement进行设值。

- StatementHandler 通过 List<E> query(Statement statement, ResultHandler resultHandler)方法来完成执行Statement，和将Statement对象返回的resultSet封装成List。

StatementHandler 的 parameterize(Statement) 方法调用了 ParameterHandler 的 setParameters(statement)方法，ParameterHandler 的 setParameters(Statement)方法负责根据输入的参数对 statement 对象的“？”占位符处进行赋值。具体源码如下：

```
public class PreparedStatementHandler extends BaseStatementHandler {
  @Override
  public void parameterize(Statement statement) throws SQLException {
    //使用 ParameterHandler 对象完成对 Statement 的设值
    parameterHandler.setParameters((PreparedStatement) statement);
  }

@Override
  public <E> List<E> query(Statement statement, ResultHandler resultHandler)
throws SQLException {
    PreparedStatement ps = (PreparedStatement) statement;
    ps.execute();
    return resultSetHandler.<E> handleResultSets(ps);
  }
}
```

从上述代码可以看出，StatementHandler 的 List<E> query(Statement statement, ResultHandler resultHandler)方法的实现，是调用了 ResultSetHandler 的 handleResultSets（Statement）方法。ResultSetHandler 的 handleResultSets(Statement) 方法会将 Statement 语句执行后生成的 resultSet 结果集转换成 List<E> 结果集。

上述步骤即 MyBatis 的大致执行流程，更多详细的内容可查看 MyBatis 源码进行学习。

9.4 思考与练习

1. MyBatis 整体框架分为哪三层？
2. 简述 SqlSession 接口。
3. 简述 MyBatis 初始化流程。
4. 简单描述 SqlSessionFactory 类的作用。
5. 简述 MyBatis 的执行流程。
6. 简述 MyBatis 核心处理层的内容。
7. MyBatis 基础支撑层包括哪几方面的任务？

第 10 章 用户管理系统项目实战

本章将综合运用之前几章讲解的内容实现一个简单的 Web 项目——用户管理系统，该项目虽然简单，但也包含了前、后端代码的开发及项目测试的内容，可以帮助大家建立一个项目开发的整体思路，大家也可以在此项目基础上进一步完善，以使该项目更有实用性。

10.1 项目概述

在企业中，经常会有员工入职/离职，通过员工管理系统可统一管理员工的信息，从而提高企业的办事效率。本项目是一个极简版的员工管理系统，包含员工入职、离职、删除员工、添加员工、更新员工信息等功能。

在开发员工管理系统之前，请读者先搭建一个 Web 项目的框架，具体操作可以参考第 2 章。

10.2 员工表设计

首先，设计员工表，具体的 SQL 代码如下：

```
CREATE TABLE `sys_user` (
  `id` bigint(32) NOT NULL COMMENT '主键',
  `name` varchar(10) DEFAULT NULL COMMENT '姓名',
  `no` varchar(10) DEFAULT NULL COMMENT '工号',
  `status` varchar(1) DEFAULT NULL COMMENT '状态，0：删除 1：在职  2：离职',
```

```
    `position` varchar(50) DEFAULT NULL COMMENT '职位',
    `reason` varchar(255) DEFAULT NULL COMMENT '离职原因'
) ENGINE=InnoDB DEFAULT CHARSET=utf8;
```

员工表主要有姓名，工号，职位以及状态（0：删除；1：在职；2：离职）等字段。将上述的 SQL 代码执行到 MySQL 数据库。同时，准备初始化的数据，具体代码如下：

```
    INSERT INTO `springmvc-mybatis-book`.`sys_user` (`id`, `name`, `no`,
`status`, `position`) VALUES ('1', 'ay', '001', '1', 'java 工程师');
    INSERT INTO `springmvc-mybatis-book`.`sys_user` (`id`, `name`, `no`,
`status`, `position`) VALUES ('2', 'al', '002', '2', '产品经理');
```

10.3 持久化类的开发

员工表设计完成之后，在 src\main\java\com\ay\model 目录下开发对应的持久化类，类名为 SysUser，具体代码如下：

```
package com.ay.model;
import java.io.Serializable;

/**
 * 员工表
 * Created by Ay on 2020/3/22
 */
public class SysUser implements Serializable{
    //主键
    private Integer id;
    //用户名
    private String name;
    //工号
    private String no;
    //职位
    private String position;
    //离职原因
    private String reason;
    //状态，0：删除 1：在职  2：离职
    private String status;

    public Integer getId() {
        return id;
    }

    public void setId(Integer id) {
        this.id = id;
```

```
    }

    public String getName() {
        return name;
    }

    public void setName(String name) {
        this.name = name;
    }

    public String getNo() {
        return no;
    }

    public void setNo(String no) {
        this.no = no;
    }

    public String getPosition() {
        return position;
    }

    public void setPosition(String position) {
        this.position = position;
    }

    public String getStatus() {
        return status;
    }

    public void setStatus(String status) {
        this.status = status;
    }

    public String getReason() {
        return reason;
    }

    public void setReason(String reason) {
        this.reason = reason;
    }
}
```

10.4　DAO 层和 Mapper 映射文件

SysUser 类开发完成之后，继续创建对应的 Dao 类和 Mapper 映射文件。在 src\main\java\com\ay\dao 目录下创建 SysUserDao.java 类，具体代码如下：

```
package com.ay.dao;
import com.ay.model.SysUser;
import org.springframework.stereotype.Repository;

import java.util.List;

/**
 * Dao 层
 * Created by Ay on 2020/3/22.
 */
@Repository
public interface SysUserDao {
    /**
     * 查询所有用户
     * @return
     */
    List<SysUser> findAll();
}
```

SysUserDao 接口只有一个 findAll 方法，用来查询所有状态为未删除的用户。

在 src\main\resources\mapper 目录下创建 SysUserMapper.xml 文件，具体代码如下：

```
<?xml version="1.0" encoding="UTF-8" ?>
<!DOCTYPE mapper PUBLIC "-//mybatis.org//DTD Mapper 3.0//EN"
        "http://mybatis.org/dtd/mybatis-3-mapper.dtd">
<mapper namespace="com.ay.dao.SysUserDao">
    <cache/>
    <resultMap id="userMap" type="com.ay.model.SysUser">
        <id property="id" column="id"/>
        <result property="name" column="name"/>
        <result property="no" column="no"/>
        <result property="position" column="position"/>
        <result property="status" column="status"/>
        <result property="reason" column="reason"/>
    </resultMap>
    <!-- sql 代码块 -->
    <sql id="table_column">
        id,
```

```
        name,
        no,
        status,
        position,
        reason
    </sql>

    <select id="findAll" resultMap="userMap">
        select
        <include refid="table_column"/>
        from sys_user
        where status != '0';
    </select>
</mapper>
```

SysUserMapper.xml 文件中，有一个 id 为 findAll 的<select>标签，findAll 即 SysUserDao 类中的 findAll 方法名，<select>标签返回结果类型为 userMap，每个结果类型都是 SysUser。<select>标签还使用<include>标签，来复用 SQL 片段。<include>标签的 refid 属性值与<sql>标签的 id 属性值一致。

10.5 接口和实现类开发

Dao 层的代码开发完成之后，接下来开发 Service 服务层接口及对应的实现类，在 src\main\java\com\ay\service 目录下创建 SysUserService.java 接口类，具体代码如下：

```
package com.ay.service;
import com.ay.model.SysUser;
import java.util.List;
/**
 * 用户服务接口
 * Created by Ay on 2020/3/22
 */
public interface SysUserService {

    List<SysUser> findAll();
}
```

在 src\main\java\com\ay\service\impl 目录下创建 SysUserServiceImpl.java 实现类，具体代码如下：

```
package com.ay.service.impl;
import com.ay.dao.SysUserDao;
import com.ay.model.SysUser;
import com.ay.service.SysUserService;
```

```
import org.springframework.stereotype.Service;
import javax.annotation.Resource;
import java.util.List;

/**
 * 用户服务实现类
 * Created by Ay on 2020/3/22
 */
@Service
public class SysUserServiceImpl implements SysUserService{
    @Resource
    private SysUserDao sysUserDao;

    public List<SysUser> findAll() {
        return sysUserDao.findAll();
    }
}
```

在 SysUserServiceImpl.java 类中，实现 SysUserService 定义的 findAll 接口，同时使用@Resource 注解注入 SysUserDao 实例。

10.6　控制层和 DTO 类的开发

服务层接口及实现类开发完成，我们继续开发控制层类，在 src\main\java\com\ay\controller 目录下创建 SysUserController.java 类，具体代码如下：

```
package com.ay.controller;
import com.ay.dto.SysUserDTO;
import com.ay.model.SysUser;
import com.ay.service.SysUserService;
import org.springframework.stereotype.Controller;
import org.springframework.ui.Model;
import org.springframework.util.CollectionUtils;
import org.springframework.web.bind.annotation.GetMapping;
import org.springframework.web.bind.annotation.RequestMapping;
import javax.annotation.Resource;
import java.util.ArrayList;
import java.util.Collections;
import java.util.List;

/**
 * 用户控制层
 * Created by Ay on 2020/3/22
```

```
 */
@Controller
@RequestMapping("/sysUser")
public class SysUserController {

    @Resource
    private SysUserService sysUserService;

    @GetMapping("/findAll")
    public String findAll(Model model){
        List<SysUser> users = sysUserService.findAll();
        model.addAttribute("users", convertToList(users));
        return "userManage";
    }
    private List<SysUserDTO> convertToList(List<SysUser> users){
        if(CollectionUtils.isEmpty(users)){
            return Collections.EMPTY_LIST;
        }
        List<SysUserDTO> userList = new ArrayList<SysUserDTO>();
        for(SysUser user: users){
            userList.add(convertToDTO(user));
        }
        return userList;
    }

    private SysUserDTO convertToDTO(SysUser user){
        SysUserDTO userDTO = new SysUserDTO();
        userDTO.setId(String.valueOf(user.getId()));
        userDTO.setName(user.getName());
        userDTO.setNo(user.getNo());
        userDTO.setPosition(user.getPosition());
        if("1".equals(user.getStatus())){
            userDTO.setStatus("在职");
        }
        if("2".equals(user.getStatus())){
            userDTO.setStatus("已离职");
        }
        userDTO.setReason(user.getReason());
        return userDTO;
    }
}
```

在 SysUserController 控制层类中，使用@RequestMapping 和@GetMapping 注解定义请求路径：/sysUser + findAll，同时使用@Resource 注解注入 SysUserService 实例，sysUserService 调用 findAll 方法请求所有的用户，并将用户数据存入 Model 对象中，key 为

users， value 为用户列表，最后返回 userManage 字符串，该字符串代表前端名为 userManage.jsp 页面。

SysUserController 类还开发了 convertToList 方法，用户将 Model 对象转换为 DTO 对象，例如 stauts 状态为 1 的转换为“在职”，status 状态为 2 的转换为“离职”。

DTO 类代码如下所示：

```
package com.ay.dto;
import java.io.Serializable;
/**
 * 用户 DTO
 * Created by Ay on 2020/3/22.
 */
public class SysUserDTO implements Serializable{

    //主键
    private String id;
    //用户名
    private String name;
    //工号
    private String no;
    //职位
    private String position;
    //离职原因
    private String reason;
    //状态，0：删除 1：在职  2：离职
    private String status;
    //省略 set、get 方法
}
```

10.7 前端页面开发

后端代码开发完成后，在 src\main\webapp\WEB-INF\views 目录下开发 userManage.jsp 页面，具体代码如下：

```
<%@page language="java" contentType="text/html; charset=UTF-8"
       pageEncoding="UTF-8" isELIgnored="false"%>
<%@ taglib uri="http://java.sun.com/jsp/jstl/core" prefix="c"%>
<%@ taglib prefix="fmt" uri="http://java.sun.com/jstl/fmt" %>
<!DOCTYPE HTML>
<html>
<head>
    <title>Getting Started: Serving Web Content</title>
    <meta http-equiv="Content-Type" content="text/html; charset=UTF-8" />
```

```
        <script
src="https://cdn.staticfile.org/jquery/1.10.2/jquery.min.js"></script>
    </head>
    <body>
    <div align="center">用户管理系统</div><br>
    <div style="margin-left:63%"><button id="in" onclick="into()">办理入职
</button></div>
    <table border="1" align="center">
        <tr>
            <td>姓名</td>
            <td>工号</td>
            <td>职位</td>
            <td>状态</td>
            <td>操作</td>
        </tr>
    <c:forEach items="${users}" var="user">
        <tr>
            <td width="50">${user.name}</td>
            <td width="50">${user.no}</td>
            <td width="150">${user.position}</td>
            <td  width="100"> ${user.status}</td>
            <td  width="200"><button id="out" onclick="out(${user.id})">办理离职
</button>
                <button id="update" onclick="info(${user.id})">信息更新
</button></td>
        </tr>

    </c:forEach>

    </table>
    </body>
    <script>

        function into() {
            var url = "http://localhost:8080/sysUser/in"
            window.open(url,'_self');
        }

        function out(id) {
            var userId = id
            var url = "http://localhost:8080/sysUser/out?id=" + userId;
            window.open(url,'_self');
        }

        function info(id) {
```

```
        var userId = id
        var url = "http://localhost:8080/sysUser/updateInfo?id=" + userId;
        window.open(url,'_self');
    }
</script>

</html>
```

在 userManage.jsp 页面中，引入 jstl 标签库和函数库，具体代码如下：

```
<%@ taglib uri="http://java.sun.com/jsp/jstl/core" prefix="c"%>
<%@ taglib prefix="fmt" uri="http://java.sun.com/jstl/fmt" %>
```

使用<c:forEach/>标签获取后端返回的用户列表，循环遍历出每个用户，展示到 table 表格里。所有代码开发完成后，启动 tomcat 服务器，在浏览器中输入访问地址：http://localhost:8080/sysUser/findAll，便可以看到如图 10-1 所示的界面。

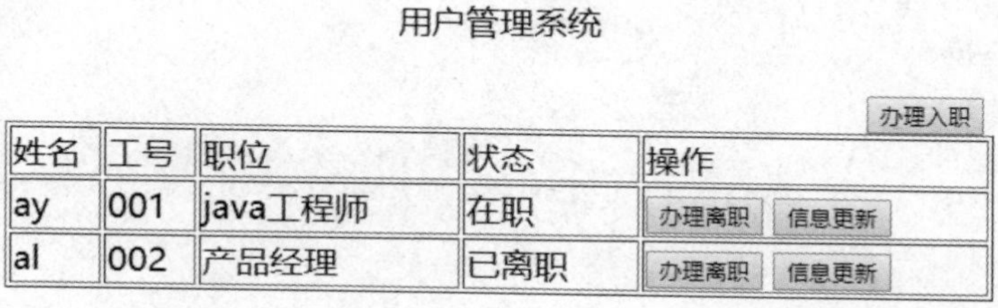

用户管理系统

办理入职

姓名	工号	职位	状态	操作
ay	001	java工程师	在职	办理离职 信息更新
al	002	产品经理	已离职	办理离职 信息更新

图 10-1 用户管理系统

10.8 员工入职/离职/更新功能

上一节，我们已经实现了查询所有用户的功能，接下来开发员工入职和离职功能。在一家大型互联网公司，入职和离职的频率比较高，员工入职实际上就是往数据库插入一条数据，状态为入职状态，员工离职实际上只要将用户的状态更改为离职状态即可。

在 src\main\resources\mapper 目录下的 SysUserMapper 代码中添加如下代码：

```
<select id="findById" resultType="com.ay.model.SysUser">
      select
      <include refid="table_column"/>
      from sys_user
      where id = #{id};
   </select>

   <insert id="save" useGeneratedKeys="true" keyProperty="id"
         parameterType="com.ay.model.SysUser">
      <selectKey keyProperty="id" resultType="int" order="BEFORE">
```

```
        select max(id) + 1 as id from sys_user
    </selectKey>
    insert into sys_user(id, name ,no, status, position)
    value (#{id},#{name},#{no},#{status},#{position} )
</insert>

<insert id="update" parameterType="com.ay.model.SysUser">
    update sys_user
    <set>
        name = #{user.name},
        no = #{user.no},
        status = #{user.status},
        position = #{user.position},
        reason = #{user.reason}
    </set>
    <where>
        id = #{user.id}
    </where>
</insert>

<insert id="updateStatus" parameterType="com.ay.model.SysUser">
    update sys_user
    <set>
        status = #{user.status},
        reason = #{user.reason}
    </set>
    <where>
        id = #{user.id}
    </where>
</insert>
```

上述代码中，findById 方法用来通过 Id 查询用户，save 方法用来插入一条数据，update 方法用来更新用户的信息，updateStatus 用来更新用户状态（在职或已离职）。

在 src\main\java\com\ay\dao 目录下的 SysUserDao 类中添加如下代码：

```
//保存用户
void save(SysUser user);
//通过用户 id 查询用户
SysUser findById(Integer id);
//更新用户信息
void update(@Param("user") SysUser user);
//更新状态（在职/离职）
void updateStatus(@Param("user") SysUser user);
```

SysUserDao 接口提供和 Mapper 文件对应的 save、findById、update 等方法。

在 src\main\java\com\ay\service 目录下的 SysUserService.java 接口添加如下代码：

```
boolean save(SysUserDTO user);
SysUser findById(Integer id);
boolean update(SysUserDTO user);
boolean updateStatus(SysUserDTO user);
```

在 src\main\java\com\ay\service\impl 目录下的 SysUserServiceImpl 实现类中添加如下代码：

```
//保存用户
public boolean save(SysUserDTO user) {
       user.setStatus("1");
       sysUserDao.save(convert(user));
       return true;
}
//更新用户
 public boolean update(SysUserDTO user) {
       SysUser sysUser = convert(user);
       sysUserDao.update(sysUser);
       return true;
 }
  //通过 id 查找用户
 public SysUser findById(Integer id) {
       return sysUserDao.findById(id);
 }
 // SysUserDTO 转换为 SysUser
 public SysUser convert(SysUserDTO user){
       SysUser sysUser = new SysUser();
       if(user.getId() != null){
          sysUser.setId(Integer.valueOf(user.getId()));
       }
       sysUser.setName(user.getName());
       sysUser.setNo(user.getNo());
       sysUser.setStatus(user.getStatus());
       if(user.getStatus().equals("在职")){
          sysUser.setStatus("1");
       }
       if(user.getStatus().equals("已离职")){
          sysUser.setStatus("2");
       }
       sysUser.setPosition(user.getPosition());
       sysUser.setReason(user.getReason());
       return sysUser;
 }
 //更新状态
 public boolean updateStatus(SysUserDTO user) {
       sysUserDao.updateStatus(convert(user));
```

```
        return true;
    }
```

SysUserServiceImpl 类实现了接口中的方法，convert 方法用于将 SysUserDTO 对象转化为 sysUser 对象。

在 src\main\java\com\ay\controller 目录下的 SysUserController 类中添加如下代码：

```
    //入职页面跳转
    @GetMapping("/in")
    public String userIn(Model model){
        return "userIn";
    }
    //离职页面跳转
    @GetMapping("/out")
    public String userOut(@RequestParam("id") String id, Model model){
        SysUser user = sysUserService.findById(Integer.valueOf(id));
        model.addAttribute("user", convertToDTO(user));
        return "userOut";
    }
    //更新用户信息页面跳转
    @GetMapping("/updateInfo")
    public String updateInfo(@RequestParam("id") String id, Model model){
        SysUser user = sysUserService.findById(Integer.valueOf(id));
        model.addAttribute("user", convertToDTO(user));
        return "updateUser";
    }
    //保存用户信息
    @PostMapping(value = "/save")
    public String save(@RequestBody SysUserDTO user){
        sysUserService.save(user);
        return "userManage";
    }
    //更新用户信息
    @PostMapping(value = "/update")
    public String update(@RequestBody SysUserDTO user){
        sysUserService.update(user);
        return "userManage";
    }
    //更新用户状态
    @PostMapping(value = "/updateStatus")
    public String updateStatus(@RequestBody SysUserDTO user){
        sysUserService.updateStatus(user);
        return "userManage";
    }
    //List<SysUser> 转换为 List<SysUserDTO>对象
    private List<SysUserDTO> convertToList(List<SysUser> users){
```

```
        if(CollectionUtils.isEmpty(users)){
            return Collections.EMPTY_LIST;
        }
        List<SysUserDTO> userList = new ArrayList<SysUserDTO>();
        for(SysUser user: users){
            //调用 convertToDTO 方法进行转换
            userList.add(convertToDTO(user));
        }
        return userList;
    }
    //SysUser 转换为 SysUserDTO 对象
    private SysUserDTO convertToDTO(SysUser user){
        SysUserDTO userDTO = new SysUserDTO();
        userDTO.setId(String.valueOf(user.getId()));
        userDTO.setName(user.getName());
        userDTO.setNo(user.getNo());
        userDTO.setPosition(user.getPosition());
        if("1".equals(user.getStatus())){
            userDTO.setStatus("在职");
        }
        if("2".equals(user.getStatus())){
            userDTO.setStatus("已离职");
        }
        userDTO.setReason(user.getReason());
        return userDTO;
    }
```

在 src\main\webapp\WEB-INF\views 目录下开发入职页面、离职页面和用户信息更新页面，具体如图 10-2、图 10-3 及图 10-4 所示。

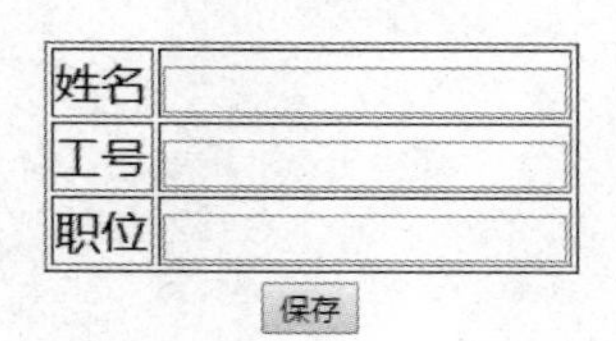

图 10-2 员工入职办理页面

员工离职办理

姓名	ay
工号	001
职位	java工程师
状态	已离职
离职原因	

确认办理

图 10-3 员工离职办理页面

员工信息更新

姓名	ay
工号	001
职位	java工程师
状态	已离职
离职原因	

确认更新

图 10-4 员工信息更新页面

员工入职办理页面的 JSP 代码如下所示：

```
<%@page language="java" contentType="text/html; charset=UTF-8"
        pageEncoding="UTF-8" isELIgnored="false"%>
<%@ taglib uri="http://java.sun.com/jsp/jstl/core" prefix="c"%>
<%@ taglib prefix="fmt" uri="http://java.sun.com/jstl/fmt" %>
```

```
    <%@ taglib prefix="form" uri="http://www.springframework.org/tags/form" %>
    <!DOCTYPE HTML>
    <html>
    <head>
       <title>Getting Started: Serving Web Content</title>
       <meta http-equiv="Content-Type" content="text/html; charset=UTF-8" />
       <script
src="https://cdn.staticfile.org/jquery/1.10.2/jquery.min.js"></script>
    </head>
    <body>
    <div align="center">新员工入职办理</div><br>
    <table border="1" align="center">
       <tr>
          <td>姓名</td><td><input id="name"/></td>
       </tr>
       <tr>
          <td>工号</td><td><input id="no"/></td>
       </tr>
       <tr>
          <td>职位</td><td><input id="position"/></td>
       </tr>
    </table>
    <div align="center"><button id="join" >保存</button></div>
    </body>
    <script>

       $("#join").click(function () {
          var url = "http://localhost:8080/sysUser/save";
          var name = $(" input[id='name' ] ").val();
          var no = $(" input[id='name' ] ").val();
          var position = $(" input[id='position' ] ").val();
          var data = {
             "name":name,
             "no":no,
             "position":position
          }
          $.ajax({
             type: "post",
             url: url,
             dataType : "text",
             contentType : "application/json",
             data: JSON.stringify(data),
             success:function (data) {
                //跳转回列表
                var url = "http://localhost:8080/sysUser/findAll"
```

```
                window.open(url,'_self');
            }
        });
    });
</script>
</html>
```

员工离职办理页面的 JSP 代码如下所示：

```
<%@page language="java" contentType="text/html; charset=UTF-8"
       pageEncoding="UTF-8" isELIgnored="false"%>
<%@ taglib uri="http://java.sun.com/jsp/jstl/core" prefix="c"%>
<%@ taglib prefix="fmt" uri="http://java.sun.com/jstl/fmt" %>
<%@ taglib prefix="form" uri="http://www.springframework.org/tags/form" %>
<!DOCTYPE HTML>
<html>
<head>
    <title>Getting Started: Serving Web Content</title>
    <meta http-equiv="Content-Type" content="text/html; charset=UTF-8" />
    <script
src="https://cdn.staticfile.org/jquery/1.10.2/jquery.min.js"></script>
</head>
<body>
<div align="center">员工离职办理</div><br>
<table border="1" align="center">
    <tr>
        <td>姓名</td><td><input id="name" value="${user.name}"/></td>
    </tr>
    <tr>
        <td>工号</td><td><input id="no" value="${user.no}"/></td>
    </tr>
    <tr>
        <td>职位</td><td><input id="position"
value="${user.position}"/></td>
    </tr>
    <tr>
        <td>状态</td><td><input id="status" value="${user.status}"/></td>
    </tr>
    <tr>
        <td>离职原因</td><td><input id="reason"></input></td>
    </tr>

</table>
<div align="center"><button id="out" value="${user.id}">确认办理
</button></div>
</body>
```

```
<script>

    $("#out").click(function () {
        var _that = this;
        var id = _that.value;
        var url = "http://localhost:8080/sysUser/updateStatus";
        var status = "2";
        var reason = $(" input[id='reason' ] ").val();
        alert(reason);
        var data = {
            "id":id,
            "status":status,
            "reason":reason
        }
        $.ajax({
            type: "post",
            url: url,
            dataType : "text",
            contentType : "application/json",
            data: JSON.stringify(data),
            success:function (data) {
                //跳转回列表
                var url = "http://localhost:8080/sysUser/findAll"
                window.open(url,'_self');
            }
        });
    });
</script>
</html>
```

员工信息更新页面对应的 JSP 代码如下所示：

```
<%@page language="java" contentType="text/html; charset=UTF-8"
        pageEncoding="UTF-8" isELIgnored="false"%>
<%@ taglib uri="http://java.sun.com/jsp/jstl/core" prefix="c"%>
<%@ taglib prefix="fmt" uri="http://java.sun.com/jstl/fmt" %>
<%@ taglib prefix="form" uri="http://www.springframework.org/tags/form" %>
<!DOCTYPE HTML>
<html>
<head>
    <title>Getting Started: Serving Web Content</title>
    <meta http-equiv="Content-Type" content="text/html; charset=UTF-8" />
    <script
src="https://cdn.staticfile.org/jquery/1.10.2/jquery.min.js"></script>
</head>
<body>
```

```
    <div align="center">员工信息更新</div><br>
    <table border="1" align="center">
        <tr>
            <td>姓名</td><td><input id="name" value="${user.name}"/></td>
        </tr>
        <tr>
            <td>工号</td><td><input id="no" value="${user.no}"/></td>
        </tr>
        <tr>
            <td>职位</td><td><input id="position"
value="${user.position}"/></td>
        </tr>
        <tr>
            <td>状态</td><td><input id="status" value="${user.status}"/></td>
        </tr>
        <tr>
            <td>离职原因</td><td><input id="reason" value="${user.reason}"/></td>
        </tr>
    </table>
    <div align="center"><button id="info" value="${user.id}">确认更新
</button></div>
    </body>
    <script>

        $("#info").click(function () {
            var _that = this;
            var id = _that.value;
            var url = "http://localhost:8080/sysUser/update";
            var name = $(" input[id='name' ] ").val();
            var no = $(" input[id='no' ] ").val();
            var position = $(" input[id='position' ] ").val();
            var status = $(" input[id='status' ] ").val();
            var reason = $(" input[id='reason' ] ").val();
            var data = {
                "id":id,
                "name":name,
                "no":no,
                "position":position,
                "status":status,
                "reason":reason
            }
            $.ajax({
                type: "post",
                url: url,
                dataType : "text",
```

```
            contentType : "application/json",
            data: JSON.stringify(data),
            success:function (data) {
                //跳转回列表
                var url = "http://localhost:8080/sysUser/findAll"
                window.open(url,'_self');
            }
        });


    });

</script>
</html>
```

10.9 测试

所有代码开发完成后，重新启动项目，在浏览器中输入地址：http://localhost:8080/sysUser/findAll，便可以看到首页，如图 10-5 所示。

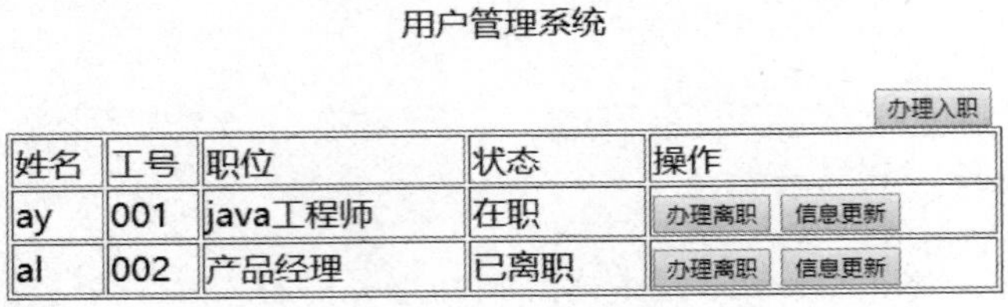

用户管理系统

办理入职

姓名	工号	职位	状态	操作
ay	001	java工程师	在职	办理离职 信息更新
al	002	产品经理	已离职	办理离职 信息更新

图 10-5　用户管理系统首页

sysUser/findAll 请求会调用 SysUserController 类的 findAll 方法，并将数据库所有的员工记录返回给前端。单击【办理入职】按钮会弹出【新员工入职办理】页面，具体如图 10-6 所示。

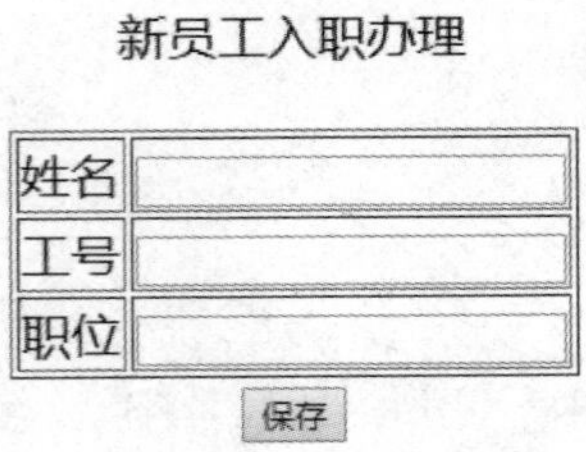

新员工入职办理

姓名	
工号	
职位	

保存

图 10-6　新员工入职办理

在【新员工入职办理】页面中，填写员工【姓名】、【工号】、【职位】等信息，单击【保存】按钮，将会触发前端 js 的 click 方法：$("#join").click(function (){}。在 click 方法中，获取表单的值，通过 Ajax 向后端发起 http://localhost:8080/sysUser/save 请求。该请求会调用 SysUserController 类的 save 方法，将数据存储到数据库。数据保存完成后，会重新请求 url：http://localhost:8080/sysUser/findAll。

单击【办理离职】按钮，将会跳转到【员工离职办理】页面，具体如图 10-7 所示。

在【员工离职办理】页面中，填写【离职原因】，单击【确认办理】后会请求 http://localhost:8080/sysUser/updateStatus，该请求会调用 SysUserController 类的 updateStatus 方法，更新当前员工的状态为离职状态，然后依然重新请求 url：http://localhost:8080/sysUser/findAll。

单击【信息更新】按钮，将会跳转到【员工信息更新】页面，如图 10-8 所示。

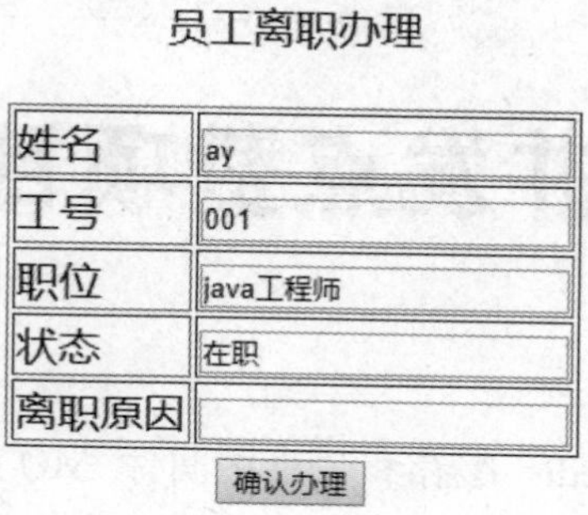

员工离职办理

姓名	ay
工号	001
职位	java工程师
状态	在职
离职原因	

确认办理

图 10-7　员工离职办理页面

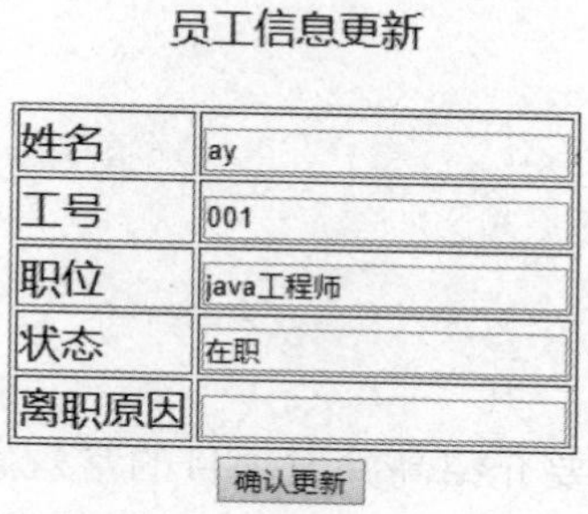

员工信息更新

姓名	ay
工号	001
职位	java工程师
状态	在职
离职原因	

确认更新

图 10-8　员工信息更新页面

在【员工信息更新】页面中，更改相关的员工信息，单击【确认更新】按钮后，会请求 http://localhost:8080/sysUser/update，该请求会调用 SysUserController 类的 update 方法更新员工的信息，然后依然重新请求 url：http://localhost:8080/sysUser/findAll。

至此，员工管理系统的开发完成。

10.10　思考与练习

1. 自己动手实现员工管理系统分页功能。
2. 在员工入职页面添加姓名长度校验和工号唯一性校验。

第11章 高并发点赞项目实战

本章主要介绍高并发项目的常规解决方案，Redis 缓存和消息中间件 MQ 的安装和使用，以及如何一步一步实现高并发点赞项目。

11.1 高并发点赞项目代码实现

11.1.1 项目概述

在社交网站或者 App 中，点赞场景非常多，比如微信说说点赞、微博点赞、博文点赞、抖音点赞等。普通人发的微信说说点赞人数比较少，所以并发数比较少；而一些名人发的微博，由于粉丝多，可能一条微博在短时间内点赞数高达 100 万甚至更多。面对如此高并发的点赞，如果项目设计没有做好，必将导致后端服务器和数据库由于压力过大程序出现异常。所以说，小小的点赞功能并不简单。大型互联网公司后端架构必会采取很多措施来解决这种高并发场景引发的问题，比如引入缓存提升读的性能、使用 MQ 队列进行异步处理等。

本章要讲的高并发点赞项目没有复杂的前端页面和业务，更多的是根据点赞这个小功能，为大家深度剖析大型互联网公司是如何解决高并发访问场景的，把这个小功能揉开掰碎展现给读者，让读者学会处理高并发场景下的架构设计，并应用到之后的工作中。

11.1.2 数据库表和持久化类

首先，根据高并发点赞功能设计数据库表 user（用户表）、mood（说说表）和 user_mood_praise_rel（点赞关联表），具体如图 11-1 所示。

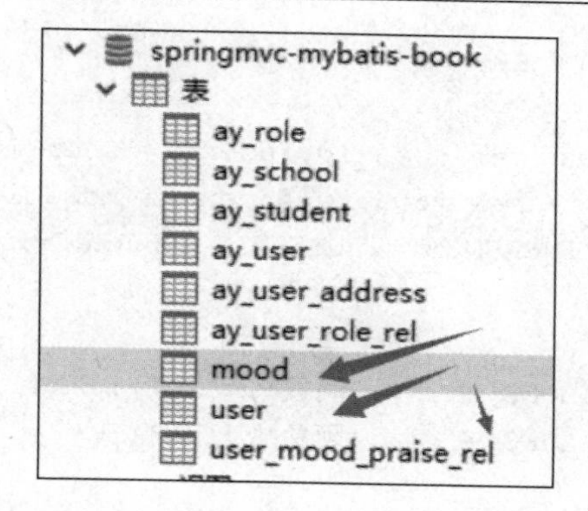

图 11-1 高并发点赞数据库表

创建 user 表（用户表）的 SQL 语句如下所示：

```
-- ----------------------------
-- Table structure for user
-- ----------------------------
DROP TABLE IF EXISTS 'user';
CREATE TABLE 'user' (
  'id' varchar(32) NOT NULL,
  'name' varchar(20) DEFAULT NULL,
  'account' varchar(20) DEFAULT NULL,
  PRIMARY KEY ('id'),
  KEY 'user_name_index' ('name') USING BTREE,
  KEY 'user_account_index' ('account') USING BTREE
) ENGINE=InnoDB DEFAULT CHARSET=utf8;
```

创建 mood 表（说说表）的 SQL 语句如下所示：

```
-- ----------------------------
-- Table structure for mood
-- ----------------------------
DROP TABLE IF EXISTS 'mood';
CREATE TABLE 'mood' (
  'id' varchar(32) NOT NULL,
  'content' varchar(256) DEFAULT NULL,
  'user_id' varchar(32) DEFAULT NULL,
  'publish_time' datetime DEFAULT NULL,
  'praise_num' int(11) DEFAULT NULL,
  PRIMARY KEY ('id'),
  KEY 'mood_user_id_index' ('user_id') USING BTREE
) ENGINE=InnoDB DEFAULT CHARSET=utf8;
```

创建 user_mood_praise_rel 表（点赞关联表）的 SQL 语句如下所示：

```
CREATE TABLE 'user_mood_praise_rel' (
  'id' bigint(32) NOT NULL AUTO_INCREMENT,
  'user_id' varchar(32) DEFAULT NULL,
```

```
    'mood_id' varchar(32) DEFAULT NULL,
    PRIMARY KEY ('id'),
    KEY 'user_mood_praise_rel_user_id_index' ('user_id') USING BTREE,
    KEY 'user_mood_praise_rel_mood_id_index' ('mood_id') USING BTREE
  ) ENGINE=InnoDB AUTO_INCREMENT=2 DEFAULT CHARSET=utf8;
```

user 表的初始化数据：

```
  INSERT INTO 'user' VALUES ('1', '阿毅', 'ay');
  INSERT INTO 'user' VALUES ('2', '阿兰', 'al');
```

mood 表的初始化数据：

```
  INSERT INTO 'mood' VALUES ('1', '今天天气真好', '1', '2018-06-30 22:09:06',
'100');
  INSERT INTO 'mood' VALUES ('2', '厦门真美，么么哒！', '2', '2018-07-29
17:13:04', '99');
```

其中，user（用户表）和 mood（说说表）是一对多的关系，即一个用户可以发表多条说说，每条说说只能属于一个用户。而 user_mood_praise_rel（点赞关联表）主要记录 user（用户表）和 mood（说说表）的关联关系，即哪些说说被哪些用户点赞。表之间的关系如图 11-2 所示。

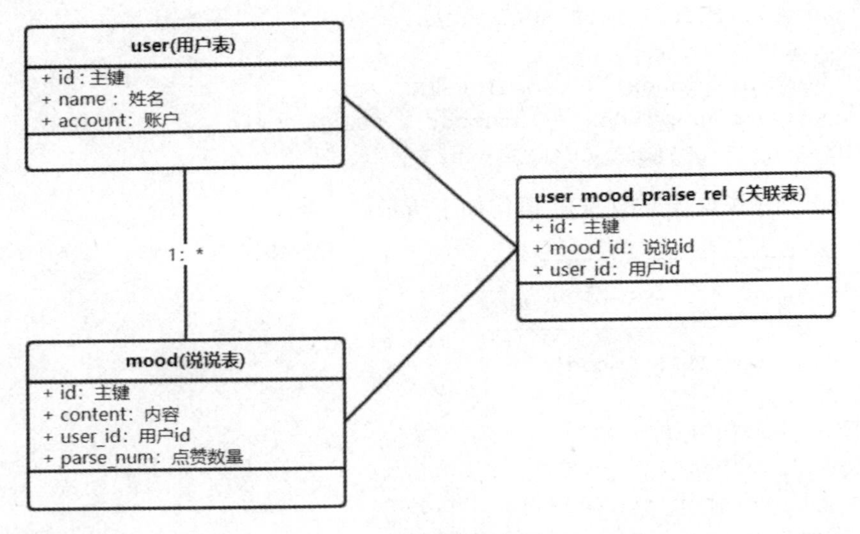

图 11-2　数据库表关系图

数据库表创建完成之后，在 springmvc-mybatis-book 项目中创建表对应的持久化对象。在\src\main\java\com.ay.model 包下创建 user 表的持久化对象，具体代码如下：

```
/**
 * 描述：用户表
 * @author Ay
 * @date 2017/9/16
 */
```

```
public class User implements Serializable {
    /**
     * 主键
     */
    private String id;
    /**
     * 用户名称
     */
    private String name;
    /**
     * 账户
     */
    private String account;

    //省略 set、get 方法
}
```

在 src\main\java\com.ay.model 包下创建 mood 表的持久化对象，具体代码如下：

```
/**
 * 描述：说说
 * @author Ay
* @date 2017/9/16
 */
public class Mood implements Serializable{
    /**
     * 主键
     */
    private String id;
    /**
     * 说说内容
     */
    private String content;
    /**
     * 点赞数量
     */
    private Integer praiseNum;
    /**
     * 用户 id
     */
    private String userId;
    /**
     * 发表时间
     */
    private Date publishTime;
```

```
    //省略 set、get 方法
}
```

在 src\main\java\com.ay.model 包下创建 user_mood_praise_rel 表的持久化对象，具体代码如下：

```
/**
 * 描述：说说点赞关联表
 *
 * @author ay
 * @date 2017/9/16
 */
public class UserMoodPraiseRel implements Serializable {
    /**
     * 主键
     */
    private String id;
    /**
     * 用户 id
     */
    private String userId;
    /**
     * 说说 id
     */
    private String moodId;

        //省略 set、get 方法
}
```

11.1.3 DAO 层和 Mapper 映射文件

数据库表和持久化类创建完成之后，继续创建对应的 Dao 类和 Mapper 映射文件。在 src\main\java\com.ay. dao 包下创建 user 表对应的 Dao 类，具体代码如下：

```
/**
 * 描述：用户 Dao
 * @author Ay
 * @create 2018/06/30
 **/
@Repository
public interface UserDao {
    //查询用户
    User find(String id);
}
```

在 src\main\resources\mapper 目录下创建 user 表对应的 Mapper 映射文件 UserMapper.xml，具体代码如下：

```
<?xml version="1.0" encoding="UTF-8" ?>
<!DOCTYPE mapper PUBLIC "-//mybatis.org//DTD Mapper 3.0//EN"
        "http://mybatis.org/dtd/mybatis-3-mapper.dtd">
<mapper namespace="com.ay.dao.UserDao">

    <resultMap id="userMap" type="com.ay.model.User">
        <id property="id" column="id"/>
        <result property="name" column="name"/>
        <result property="account" column="account"/>
    </resultMap>
    <!-- sql 代码块 -->
    <sql id="table_column">
        id,
        name,
        account
    </sql>

    <select id="find" resultMap="userMap">
        SELECT
        <include refid="table_column"/>
        FROM user
        <where>
            id = #{id}
        </where>
    </select>

</mapper>
```

UserDao 类中定义了一个查询用户的方法 find()，而 UserMapper.xml 映射配置文件中定义了 find()方法对应的 select 查询语句，还定义了 SQL 代码块。

UserDao.java 和 UserMapper.xml 文件创建完成之后，接下来创建 mood 表对应的 Dao 层和 Mapper 映射文件。

src\main\java\com.ay.dao\MoodDao.java 具体代码如下：

```
/**
 * @author Ay
 * @create 2018/06/30
 **/
@Repository
public interface MoodDao {

    List<Mood> findAll();
}
```

src\main\resources\mapper\ MoodMapper.xml 具体代码如下：

```
<?xml version="1.0" encoding="UTF-8" ?>
<!DOCTYPE mapper PUBLIC "-//mybatis.org//DTD Mapper 3.0//EN"
        "http://mybatis.org/dtd/mybatis-3-mapper.dtd">
<mapper namespace="com.ay.dao.MoodDao">

    <resultMap id="moodMap" type="com.ay.model.Mood">
        <id property="id" column="id"/>
        <result property="content" column="content"/>
        <result property="userId" column="user_id"/>
        <result property="praiseNum" column="praise_num"/>
        <result property="publishTime" column="publish_time"/>
    </resultMap>

    <sql id="table_column">
        id,
        content,
        user_id,
        praise_num,
        publish_time
    </sql>

    <select id="findAll" resultMap="moodMap">
        SELECT
        <include refid="table_column"/>
        FROM mood
    </select>

</mapper>
```

MoodDao 类中定义了一个查询所有用户的方法 findAll()，而 MoodMapper.xml 映射配置文件中定义了 findAll()方法对应的 select 查询语句，还定义了 SQL 代码块。

src\main\java\com\ay\dao\UserMoodPraiseRelDao.java 具体代码如下：

```
/**
 * 描述：用户说说点赞关联 DAO
 * @author Ay
 * @create 2018/07/01
 **/
@Repository
public interface UserMoodPraiseRelDao {

    boolean save(@Param("userMoodPraiseRel") UserMoodPraiseRel
userMoodPraiseRel);
}
```

src\main\resources\mapper\UserMoodPraiseRelMapper.xml 具体代码如下所示：

```
<?xml version="1.0" encoding="UTF-8" ?>
<!DOCTYPE mapper PUBLIC "-//mybatis.org//DTD Mapper 3.0//EN"
        "http://mybatis.org/dtd/mybatis-3-mapper.dtd">
<mapper namespace="com.ay.dao.UserMoodPraiseRelDao">
    <insert id="save" useGeneratedKeys="true" keyProperty="id"
            parameterType="com.ay.model.UserMoodPraiseRel">
        insert into user_mood_praise_rel (user_id, mood_id)
        VALUE (#{userMoodPraiseRel.userId}, #{userMoodPraiseRel.moodId})
    </insert>
</mapper>
```

UserMoodPraiseRelDao 接口只有一个 save()方法用来保存说说被用户点赞的记录。

至此，Dao 类和 Mapper 映射文件全部创建完成。代码相对比较简单，就不做过多的解释。

11.1.4　Service 层和 DTO 类

上一节中，已经在 springmvc-mybatis-book 项目中创建了 Dao 类和 Mapper 映射文件，这一节主要创建 user 表和 mood 表对应的 Service 服务层以及 DTO 对象。

src\main\java\com.ay.dto\MoodDTO.java 对应的代码如下：

```
/**
 * 描述：说说 DTO
 * @author Ay
 * @date 2018/1/6
 */
public class MoodDTO extends Mood {
    /**
     * 用户名称
     */
    private String userName;

    /**
     * 用户的账号
     */
    private String userAccount;

        //省略 set、get 方法
}
```

src\main\java\com.ay.dto\UserDTO.java 对应的代码如下：

```
/**
 * 描述：用户 DTO
 * @author Ay
 * @create 2018/07/01
```

```
 **/
public class UserDTO extends User {

}
```

MoodDTO 和 UserDTO 主要用于前端展示用的 DTO 对象，内容比较简单，不做过多的描述。

src\main\java\com.ay.service\UserService.java 对应的代码如下：

```
/**
 * 描述：用户服务接口
 * @author Ay
 * @date 2018/1/6
 */
public interface UserService {
    //通过 id 查询用户
    UserDTO find(String id);
}
```

src\main\java\com.ay.service\MoodService.java 对应的代码如下：

```
/**
 * 描述：说说接口
 * @author Ay
 * @date 2018/1/6
 */
public interface MoodService {
    //查询所有的说说
    List<MoodDTO> findAll();
}
```

UserService 用户服务层只有一个方法 find()，用来查询用户。MoodService 说说服务层定义各 findAll()方法，用来查询所有的说说。内容比较简单，不做过多的描述。

src\main\java\com.ay.service.impl\UserServiceImpl.java 对应的代码如下：

```
/**
 * 描述：用户服务类
 * @author Ay
 * @date 2018/1/6
 */
@Service
public class UserServiceImpl implements UserService {

    @Resource
    private UserDao userDao;

    public UserDTO find(String id) {
```

```
        User user =  userDao.find(id);
        return converModel2DTO(user);
    }

    private UserDTO converModel2DTO(User user){
        UserDTO userDTO = new UserDTO();
        userDTO.setId(user.getId());
        userDTO.setAccount(user.getAccount());
        userDTO.setName(user.getName());
        return userDTO;
    }
}
```

UserServiceImpl 用户服务实现类主要是对 UserService 接口进行实现。在 UserServiceImpl 类中除了实现 find()方法，还定义了私有方法用来把 User 对象转换为 UserDTO 对象。

src\main\java\com.ay.service.impl\MoodServiveImpl.java 对应的代码如下：

```
/**
 * 描述：说说服务类
 * @author Ay
 * @date 2018/1/6
 */
@Service
public class MoodServiveImpl implements MoodService {
    @Resource
    private MoodDao moodDao;

    @Resource
    private UserDao userDao;

    public List<MoodDTO> findAll() {
        //查询所有的说说
        List<Mood> moodList = moodDao.findAll();
        //转换为 DTO 对象
        return converModel2DTO(moodList);
    }

    private List<MoodDTO> converModel2DTO(List<Mood> moodList){
        if(CollectionUtils.isEmpty(moodList)) return Collections.EMPTY_LIST;
        List<MoodDTO> moodDTOList = new ArrayList<MoodDTO>();
        for(Mood mood:moodList){
            MoodDTO moodDTO = new MoodDTO();
            moodDTO.setId(mood.getId());
            moodDTO.setContent(mood.getContent());
            moodDTO.setPraiseNum(mood.getPraiseNum());
            moodDTO.setPublishTime(mood.getPublishTime());
```

```
            moodDTO.setUserId(mood.getUserId());
            moodDTOList.add(moodDTO);
            //设置用户信息
            User user = userDao.find(mood.getUserId());
            moodDTO.setUserName(user.getName());
            moodDTO.setUserAccount(user.getAccount());
        }
        return moodDTOList;
    }
}
```

MoodServiveImpl 说说服务实现类主要是对接口 MoodServive 进行实现。在 MoodServiveImpl 类中除了实现 findAll()方法外，还定义了私有方法 converModel2DTO()用来把 Mood 对象转换为 MoodDTO 对象，在方法 converModel2DTO 中还通过 userId 查询每一条说说的对应的用户，然后把用户姓名 name 和用户账号 account 设置到 MoodDTO 对象，最后返回。这样就可以清楚地知道每条说说具体是由哪个用户发表的。

src\main\java\com\ay\service\UserMoodPraiseRelService.java 具体代码如下：

```
/**
 * 描述：用户说说点赞关联接口
 * @author Ay
 * @date 2018/1/6
 */
public interface UserMoodPraiseRelService {

    boolean save(UserMoodPraiseRel userMoodPraiseRel);
}
```

src\main\java\com\ay\service\impl\UserMoodPraiseRelServiceImpl.java 具体代码如下：

```
/**
 * 描述：用户说说点赞关联服务类
 * @author Ay
 * @date  2018/1/6
 */
@Service
public class UserMoodPraiseRelServiceImpl implements
UserMoodPraiseRelService {

    @Resource
    private UserMoodPraiseRelDao userMoodPraiseRelDao;

    public boolean save(UserMoodPraiseRel userMoodPraiseRel) {
        return userMoodPraiseRelDao.save(userMoodPraiseRel);
    }
}
```

11.1.5　Controller 层和前端页面

上一节，已经创建完 Service 层和 DTO 对象，本节继续创建 user 表和 mood 表对应的 Controller 层和前端页面。

src\main\java\com.ay.controller\UserController.java 对应的代码如下：

```
/**
 * 描述：用户控制层
 * @author Ay
 * @date 2018/6/6
 */
@RestController
@RequestMapping("/user")
public class UserController {
    @Resource
    private UserService userService;
}
```

src\main\java\com.ay.controller\MoodController.java 对应的代码如下：

```
/**
 * 描述：说说控制层
 * @author Ay
 * @date 2018/1/6
 */
@Controller
@RequestMapping("/mood")
public class MoodController {

    @Resource
    private MoodService moodService;
    @RequestMapping("/findAll")
    public String findAll(Model model) {
        List<MoodDTO> moodDTOList = moodService.findAll();
        model.addAttribute("moods",moodDTOList);
        return "mood";
    }
}
```

UserController 和 MoodController 是用户和说说对应的控制层类，UserController 没有多少代码，后续需要再开发，而 MoodController 类只有一个 findAll()方法用来查询所有的说说列表。

src\main\webapp\WEB-INF\views\mood.jsp 对应的代码如下：

```
<%@page language="java" contentType="text/html; charset=UTF-8"
```

```
        pageEncoding="UTF-8" isELIgnored="false"%>
<%@ taglib uri="http://java.sun.com/jsp/jstl/core" prefix="c"%>
<!DOCTYPE HTML>
<html>
<head>
    <title>Getting Started: Serving Web Content</title>
    <meta http-equiv="Content-Type" content="text/html; charset=UTF-8" />
</head>
<body>

<div id="moods">
    <b>说说列表:</b><br>
    <c:forEach items="${moods}" var="mood">
        ------------------------------------
        <br>
        <b>用户：</b><span id="account">${mood.userName}</span><br>
        <b>说说内容：</b><span id="content">${mood.content}</span><br>
        <b>发表时间：</b>
        <span id="publish_time">
                ${mood.publishTime}
        </span><br>
        <b>点赞数：</b><span id="praise_num">${mood.praiseNum}</span><br>
        <div style="margin-left: 350px">
            <a id="praise">赞</a>
        </div>
    </c:forEach>
</div>
</body>
<script></script>
</html>
```

mood.jsp 页面主要用来展示所有的说说，并提供点赞功能。当然，这个功能还没有开发，后续内容会开发这个功能。

11.1.6 测试

所有的代码都开发完成之后，重新部署 pringmvc-mybatis-book 项目。项目启动成功之后，在浏览器输入访问路径：http://localhost:8080/mood/findAll，便可以在浏览器看到如图 11-3 所示的页面。

说说列表:

用户：阿毅

说说内容：今天天气真好

发表时间：Sun Jul 01 06:09:06 CST 2018

点赞数：100

赞

图 11-3　说说页面

11.2 传统点赞功能实现

11.2.1　概述

对于初级开发工程师，其实现点赞功能的思路如图 11-4 所示。

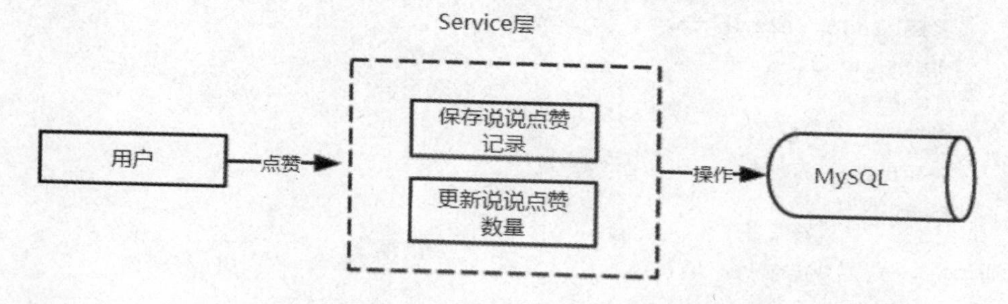

图 11-4　传统点赞功能设计方案

由图 11-4 可知，用户在前端页面（见图 11-5）给某一条喜欢的说说点赞，点赞请求到达后端时，后端开启一个线程进行处理，Service 层主要做两件事：

图 11-5　说说页面

（1）保存点赞用户与被点赞说说的关联关系，该关系保存在 user_mood_praise_rel 这张表中。

（2）更新被点赞说说的点赞数量。

Service 层处理过程中，请求数据库获取连接，执行相关的数据库操作后归还数据库连接或者杀掉数据库连接（注：取决于是否配置数据库连接池），最终返回数据给用户。

通过上述分析能够清楚地知道，在高并发情况下，比如胡歌发表一条热门说说：“今天我结婚啦”，在半个小时内说说点赞数量高达 20 万，那么 QPS 高达 111（QPS = 每秒请求数 / 事务数量），也就是后端服务每秒要创建 111 个线程来处理点赞请求，而每次请求需要创建数据库连接或者从数据库连接池获取连接。我们都知道，数据库的连接数量是有限的，这么高的线程请求数和数据库连接数对服务器来说压力非常大，会导致服务器响应时间长，处理缓慢甚至宕机。这就是传统实现所带来的隐患和弊端。

11.2.2 代码实现

分析完传统点赞功能实现的基本思路和步骤后，我们来简单地实现它。具体如下所示：

在 src\main\resources\mapper\MoodMapper.xml 文件中添加如下代码：

```
<select id="findById" resultMap="moodMap">
        select
        <include refid="table_column"/>
        from mood
        <where>
            id = #{id}
        </where>
    </select>

<update id="update">
        update mood
        <set>
            <if test="mood.content != null and mood.content != ''">
                content = #{mood.content},
            </if>
            <if test="mood.praiseNum != null and mood.praiseNum != ''">
                praise_num = #{mood.praiseNum},
            </if>
        </set>
        WHERE id = #{mood.id}
 </update>
```

findById 查询用来根据 id 查询说说实体 mood，update 方法主要用来更新说说数据。代码比较简单，就不过多赘述。

在 src\main\java\com\ay\dao\MoodDao.java 文件中添加如下代码：

```
/**
 * 描述：用户 Dao
 * @author Ay
 * @create 2018/06/30
 **/
@Repository
public interface MoodDao {
```

```
    boolean update(@Param("mood")Mood mood);
    Mood findById(String id);
}
```

在 src\main\java\com\ay\service\MoodService.java 文件中添加如下代码：

```
/**
 * 描述：说说接口
 * @author Ay
 * @date 2018/1/6
 */
public interface MoodService {
    //传统点赞
    boolean praiseMood(String userId, String moodId);
    boolean update(@Param("mood") Mood mood);
    Mood findById(String id);
}
```

MoodService 接口主要添加三个方法：praiseMood()处理用户点赞、update()更新说说数据、findById()根据 id 查询说说。

在 src\main\java\com\ay\service\impl\MoodServiveImpl.java 文件中添加如下代码：

```
/**
 * 描述：说说服务类
 * @author Ay
 * @date 2018/1/6
 */
@Service
public class MoodServiveImpl implements MoodService {
    @Resource
    private MoodDao moodDao;
    @Resource
    private UserDao userDao;
    @Resource
    private UserMoodPraiseRelDao userMoodPraiseRelDao;

    //省略代码

    public boolean praiseMood(String userId, String moodId) {
        //保存关联关系
        UserMoodPraiseRel userMoodPraiseRel = new UserMoodPraiseRel();
        userMoodPraiseRel.setUserId(userId);
        userMoodPraiseRel.setMoodId(moodId);
        userMoodPraiseRelDao.save(userMoodPraiseRel);
        //更新说说的点赞数量
        Mood mood = this.findById(moodId);
        mood.setPraiseNum(mood.getPraiseNum() + 1);
```

```
        this.update(mood);

        return Boolean.TRUE;
    }

    public boolean update(Mood mood) {
        return moodDao.update(mood);
    }

    public Mood findById(String id) {
        return moodDao.findById(id);
    }
}
```

MoodServiveImpl 实现 MoodServive 接口中的 praiseMood()、update()和 findById()方法。

在 src\main\java\com\ay\controller\MoodController.java 文件中添加如下代码：

```
/**
 * 描述：说说控制层
 * @author Ay
 * @date 2018/1/6
 */
@Controller
@RequestMapping("/mood")
public class MoodController {

    @Resource
    private MoodService moodService;

        //省略代码

    @GetMapping(value = "/{moodId}/praise")
    public String praise(Model model, @PathVariable(value="moodId")String
moodId,
                        @RequestParam(value="userId")String userId){
        boolean isPraise = moodService.praiseMood(userId, moodId);

        List<MoodDTO> moodDTOList = moodService.findAll();
        model.addAttribute("moods",moodDTOList);
        model.addAttribute("isPraise", isPraise);
        return "mood";
    }
}
```

在 src\main\webapp\WEB-INF\views\mood.jsp 文件中添加如下代码：

```
<%@page language="java" contentType="text/html; charset=UTF-8"
        pageEncoding="UTF-8" isELIgnored="false"%>
```

```
<%@ taglib uri="http://java.sun.com/jsp/jstl/core" prefix="c"%>
<%@ taglib prefix="fmt" uri="http://java.sun.com/jstl/fmt" %>
<!DOCTYPE HTML>
<html>
<head>
    <title>Getting Started: Serving Web Content</title>
    <meta http-equiv="Content-Type" content="text/html; charset=UTF-8" />
</head>
<body>

<div id="moods">
    <b>说说列表:</b><br>
    <c:forEach items="${moods}" var="mood">
        ------------------------------------
        <br>
        <b>用户：</b><span id="account">${mood.userName}</span><br>
        <b>说说内容：</b><span id="content">${mood.content}</span><br>
        <b>发表时间：</b>
        <span id="publish_time">
                ${mood.publishTime}
        </span><br>
        <b>点赞数：</b><span id="praise_num">${mood.praiseNum}</span><br>
        <div style="margin-left: 350px">
            <!-- 重点，每次点赞都会向后端发起请求，并把请求参数 moodId 和 userId 传递给
后端 -->
            <a id="praise" href="/mood/${mood.id}/praise?userId=
${mood.userId}">赞</a>
        </div>
    </c:forEach>
</div>
</body>
<script></script>
</html>
```

11.2.3　测试

代码开发完成之后，重新启动项目 springmvc-mybatis-book，在浏览器中输入请求 URL：http://localhost:8080/mood/findAll 时，后端会返回所有的说说数据给前端展示，具体如图 11-6 所示。

单击赞功能，发现说说点赞数的数量会不断地增加，同时在数据库表 user_mood_praise_rel 中，可以查询到相关的关联数据。当然这里的点赞功能允许同一个用户对同一条说说进行多次点赞，在真实的场景下是不允许的，比如微博、微信等。我们只是为了开发方便，重点是掌握解决高并发点赞功能的思路和方法。

说说列表:
用户：阿毅
说说内容：今天天气真好!
发表时间：Sun Jul 01 06:09:06 CST 2018
点赞数：117
赞
说说列表:
用户：阿毅
说说内容：今天天气真好!
发表时间：Sun Jul 01 06:09:06 CST 2018
点赞数：119
赞

图 11-6　说说列表页面

11.3　集成 Redis 缓存

11.3.1　概述

传统的点赞功能实现暴露的问题是很明显的，主要有：

（1）高并发请求下，服务器频繁创建线程。

（2）高并发请求下，数据库连接池中的连接数有限。

（3）高并发请求下，点赞功能是同步处理等。

传统的点赞功能，性能瓶颈主要在第（2）和（3）条，第（1）条问题不可避免。针对传统的点赞功能实现所带来的一系列问题，我们引入 Redis 缓存。每次点赞请求不是直接和 MySQL 数据库进行交互，而是和 Redis 缓存服务器进行交互，即把点赞相关的数据保存到 Redis 缓存，最后通过 Quartz 创建定时计划，再把缓存中的数据保存到数据库。具体解决方案如图 11-7 所示。

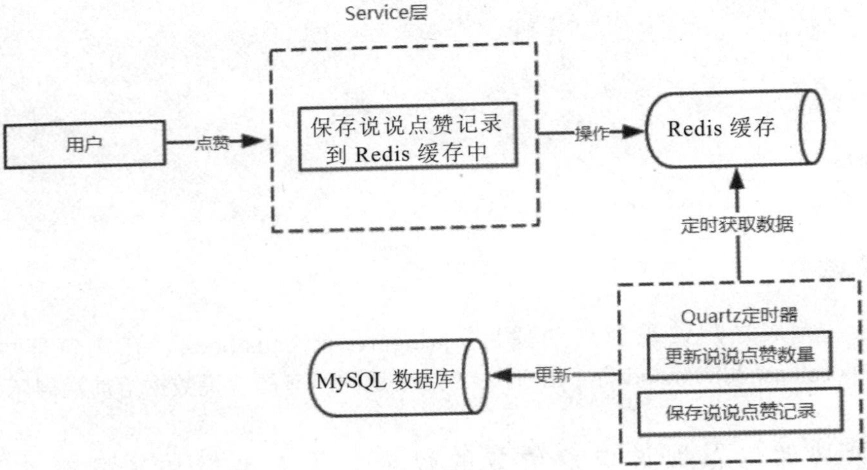
Service层
用户
点赞
保存说说点赞记录
到 Redis 缓存中
操作
Redis 缓存
定时获取数据
Quartz定时器
更新说说点赞数量
保存说说点赞记录
更新
MySQL 数据库

图 11-7　引入 Redis 缓存设计方案

11.3.2 Redis的安装和使用

Redis是一个基于内存的，单线程高性能key-value型数据库，读写性能优异。和Memcached缓存相比，Redis支持丰富的数据类型，包括string（字符串）、list（链表）、set（集合）、zset（sorted set有序集合）和hash（哈希类型）。因此Redis在企业业中被广泛使用。

Redis项目本身不支持Windows，但是Microsoft开放技术小组开发和维护这个Windows端口（针对Win64）。所以我们可以在网络上下载Redis的Windows版本。具体步骤如下：

步骤01 打开官网http://redis.io/，单击Download，具体如图11-8所示。

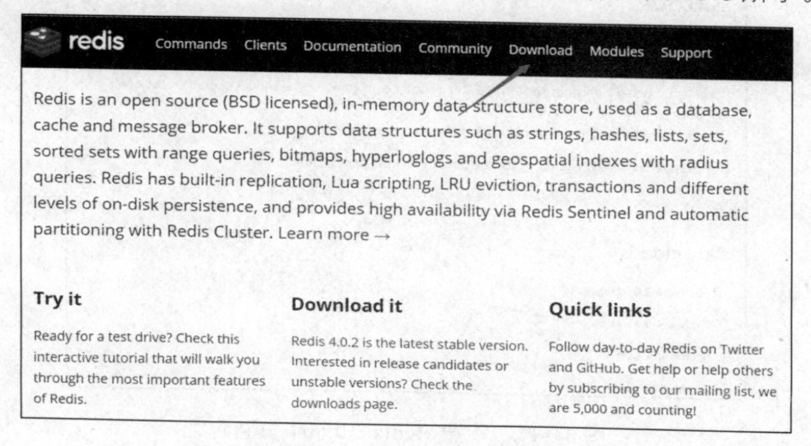

图11-8 Redis下载首页

步骤02 在弹出的页面中，找到Learn more选项，并单击进入，具体如图11-9所示。

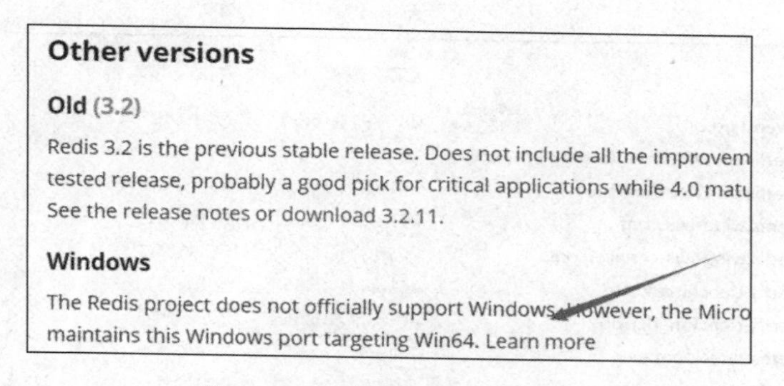

图11-9 点击Learn more链接

步骤03 在弹出的页面中选择【releases】选项，具体如图11-10所示。

步骤04 在弹出的界面中选择Redis3.0.504这个版本，选择其他版本也可以，单击【Redis-x64-3.0.504.zip】下载Redis安装包，如图11-11所示。

MicrosoftArchive / redis
forked from antirez/redis
Code　Issues 131　Pull requests 7　Projects 0　Wiki
Redis is an in-memory database that persists on disk. The data model is key-value, supported: Strings, Lists, Sets, Sorted Sets, Hashes http://redis.io
5,056 commits　20 branches　37 releases

图 11-10　选择 Download ZIP 下载 Redis

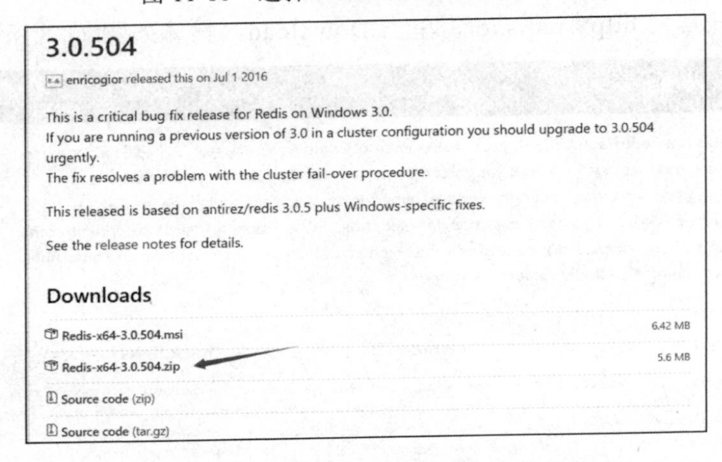

图 11-11　下载 Redis 3.0.504 安装包

步骤 05 解压下载的安装包【Redis-x64-3.0.504.zip】，双击【redis-server.exe】，Redis 服务就运行起来了，如图 11-12 所示。同时我们可以看到 Redis 启动成功的界面，如图 11-13 所示。

名称	修改日期	类型	大小
EventLog.dll	2016/7/1 15:54	应用程序扩展	1 KB
Redis on Windows Release Notes.docx	2016/7/1 15:52	Microsoft Word ...	13 KB
Redis on Windows.docx	2016/7/1 15:52	Microsoft Word ...	17 KB
redis.windows.conf	2016/7/1 15:52	CONF 文件	43 KB
redis.windows-service.conf	2016/7/1 15:52	CONF 文件	43 KB
redis-benchmark.exe	2016/7/1 15:55	应用程序	397 KB
redis-benchmark.pdb	2016/7/1 15:55	PDB 文件	4,268 KB
redis-check-aof.exe	2016/7/1 15:55	应用程序	251 KB
redis-check-aof.pdb	2016/7/1 15:55	PDB 文件	3,436 KB
redis-check-dump.exe	2016/7/1 15:55	应用程序	262 KB
redis-check-dump.pdb	2016/7/1 15:55	PDB 文件	3,404 KB
redis-cli.exe	2016/7/1 15:55	应用程序	471 KB
redis-cli.pdb	2016/7/1 15:55	PDB 文件	4,412 KB
redis-server.exe	2016/7/1 15:55	应用程序	1,517 KB
redis-server.pdb	2016/7/1 15:55	PDB 文件	6,748 KB
Windows Service Documentation.docx	2016/7/1 9:17	Microsoft Word ...	14 KB

图 11-12　启动 Redis 服务器

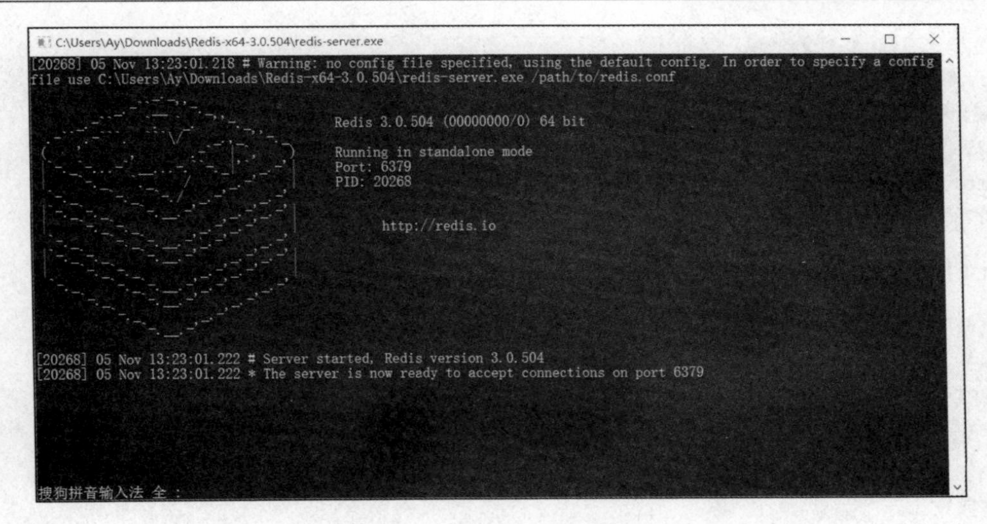

图 11-13　Redis 启动成功界面

Redis 安装成功之后，可以在安装包里找到 Redis 客户端程序 redis-cli.exe，如图 11-14 所示，双击 redis-cli.exe，打开 Redis 客户端界面，如图 11-15 所示。

redis-check-aof.exe	2016/7/1 15:55
redis-check-aof.pdb	2016/7/1 15:55
redis-check-dump.exe	2016/7/1 15:55
redis-check-dump.pdb	2016/7/1 15:55
redis-cli.exe	2016/7/1 15:55
redis-cli.pdb	2016/7/1 15:55
redis-server.exe	2016/7/1 15:55
redis-server.pdb	2016/7/1 15:55
Windows Service Documentation.docx	2016/7/1 9:17

图 11-14　启动 Redis 客户端

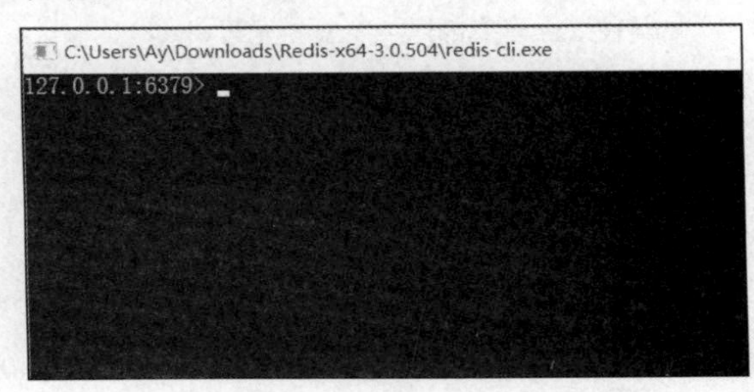

图 11-15　Redis 启动成功界面

下面就使用 Redis 客户端对 Redis 的几种数据类型做基本的增删改查操作练习，具体代码如下：

字符串类型的增删改查：

```
###增加一个值 key 为 name，value 为 ay
127.0.0.1:6379> set name 'ay'
OK
###查询 name 的值
127.0.0.1:6379> get name
"ay"
###更新 name 的值为 al
127.0.0.1:6379> set name 'al'
OK
###查询 name 的值
127.0.0.1:6379> get name
"al"
###删除 name 的值
```

```
127.0.0.1:6379> del name
(integer) 1
###查询是否存在 name，0 代表不存在
127.0.0.1:6379> exists name
(integer) 0
127.0.0.1:6379>
```

List 集合的增删改查：

```
###添加 key 为 user_list，value 为'ay','al'的 list 集合
127.0.0.1:6379> lpush user_list 'ay' 'al'
(integer) 2
###查询 key 为 user_list 的集合
127.0.0.1:6379> lrange user_list 0 -1
1) "al"
2) "ay"
###往 list 尾部添加 love 元素
127.0.0.1:6379> rpush user_list 'love'
(integer) 3
###往 list 头部添加 hope 元素
127.0.0.1:6379> lpush user_list 'hope'
(integer) 4
###查询 key 为 user_list 的集合
127.0.0.1:6379> lrange user_list 0 -1
1) "hope"
2) "al"
3) "ay"
4) "love"
###更新 index 为 0 的值
127.0.0.1:6379> lset user_list 0 'wish'
OK
###查询 key 为 user_list 的集合
127.0.0.1:6379> lrange user_list 0 -1
1) "wish"
2) "al"
3) "ay"
4) "love"
###删除 index 为 0 的值
127.0.0.1:6379> lrem user_list 0 'wish'
(integer) 1
###查询 key 为 user_list 的集合
127.0.0.1:6379> lrange user_list 0 -1
1) "al"
2) "ay"
3) "love"
127.0.0.1:6379>
```

Set 集合的增删改查：

```
###添加 key 为 user_set,value 为"ay" "al"  "love"的集合
127.0.0.1:6379> sadd user_set "ay" "al"  "love"
(integer) 3
###查询 key 为 user_set 集合
127.0.0.1:6379> smembers user_set
1) "al"
2) "ay"
3) "love"
###删除 value 为 love，返回 1 表示删除成功，0 表示失败
127.0.0.1:6379> srem user_set 'love'
(integer) 1
###查询 set 集合所有值
127.0.0.1:6379> smembers user_set
1) "al"
2) "ay"
###添加 love 元素，set 集合是没有顺序的，所以无法判断添加到那个位置
127.0.0.1:6379> sadd user_set 'love'
(integer) 1
###查询 set 集合所有值，发现添加到第二个位置
127.0.0.1:6379> smembers user_set
1) "al"
2) "love"
3) "ay"
###添加 love 元素，由于 set 集合已经存在，返回 0 代表添加不成功，但是不会报错
127.0.0.1:6379> sadd user_set 'love'
(integer) 0
```

Hash 集合的增删改查：

```
###清除数据库
127.0.0.1:6379> flushdb
OK
###创建 hash，key 为 user_hset,字段为 user1，值为 ay
127.0.0.1:6379> hset user_hset "user1"  "ay"
(integer) 1
###往 key 为 user_hset 添加字段为 user2，值为 al
127.0.0.1:6379> hset user_hset "user2"  "al"
(integer) 1
###查询 user_hset 字段长度
127.0.0.1:6379> hlen user_hset
(integer) 2
###查询 user_hset 所有字段
127.0.0.1:6379> hkeys user_hset
1) "user1"
```

```
2) "user2"
###查询 user_hset 所有值
127.0.0.1:6379> hvals user_hset
1) "ay"
2) "al"
###查询字段 user1 的值
127.0.0.1:6379> hget user_hset "user1"
"ay"
###获取 key 为 user_hset 所有的字段和值
127.0.0.1:6379> hgetall user_hset
1) "user1"
2) "ay"
3) "user2"
4) "al"
###更新字段 user1 的值为 new_ay
127.0.0.1:6379> hset user_hset "user1" "new_ay"
(integer) 0
###更新字段 user2 的值为 new_al
127.0.0.1:6379> hset user_hset "user2" "new_al"
(integer) 0
###获取 key 为 user_hset 所有的字段和值
127.0.0.1:6379> hgetall user_hset
1) "user1"
2) "new_ay"
3) "user2"
4) "new_al"
###删除字段 user1 和值
127.0.0.1:6379> hdel user_hset user1
(integer) 1
###获取 key 为 user_hset 所有的字段和值
127.0.0.1:6379> hgetall user_hset
1) "user2"
2) "new_al"
127.0.0.1:6379>
```

SortedSet 集合的增删改查：

```
###清除数据库
127.0.0.1:6379> flushdb
OK
###SortedSet 集合添加 ay 元素，分数为 1
127.0.0.1:6379> zadd user_zset  1 "ay"
(integer) 1
###SortedSet 集合添加 al 元素，分数为 2
127.0.0.1:6379> zadd user_zset  2 "al"
(integer) 1
```

```
###SortedSet 集合添加 love 元素，分数为 3
127.0.0.1:6379> zadd user_zset  3 "love"
(integer) 1
###按照分数由小到大查询 user_zset 集合元素
127.0.0.1:6379> zrange user_zset 0 -1
1) "ay"
2) "al"
3) "love"
###按照分数由大到小查询 user_zset 集合元素
127.0.0.1:6379> zrevrange user_zset 0 -1
1) "love"
2) "al"
3) "ay"
###查询元素 ay 的分数值
127.0.0.1:6379> zscore user_zset "ay"
"1"
###查询元素 love 的分数值
127.0.0.1:6379> zscore user_zset "love"
"3"
```

11.3.3　集成 Redis 缓存

在 SSM 框架中集成 Redis 缓存，首先需要在 pom.xml 文件中引入所需的依赖，具体代码如下：

```
<properties>
   <spring.redis.version>1.6.0.RELEASE</spring.redis.version>
   <jedis.version>2.7.2</jedis.version>
   <commons.version>2.4.2</commons.version>
</properties>
<!-- 集成 redis -->
<dependency>
     <groupId>org.springframework.data</groupId>
     <artifactId>spring-data-redis</artifactId>
     <version>${spring.redis.version}</version>
   </dependency>
 <dependency>
     <groupId>org.apache.commons</groupId>
     <artifactId>commons-pool2</artifactId>
     <version>${commons.version}</version>
 </dependency>
 <dependency>
     <groupId>redis.clients</groupId>
     <artifactId>jedis</artifactId>
     <version>${jedis.version}</version>
```

```
    </dependency>
```

在 pom 文件引入 Redis 所需的依赖之后，还需要在 src\main\resources 目录下创建 redis.properties 文件并添加如下的配置信息：

```
redis.maxIdle=300
redis.minIdle=100
redis.maxWaitMillis=3000
redis.testOnBorrow=true
redis.maxTotal=500
//服务器 id 地址
redis.host=127.0.0.1
//链接端口，默认 6379
redis.port=6379
//redis 密码默认为空
redis.password=
```

在 src\main\resources 目录下创建 spring-redis.xml 配置文件并添加如下配置信息：

```
<?xml version="1.0" encoding="UTF-8"?>
<beans   xmlns="http://www.springframework.org/schema/beans"
        xmlns:xsi="http://www.w3.org/2001/XMLSchema-instance"
        xmlns:p="http://www.springframework.org/schema/p"
        xmlns:tx="http://www.springframework.org/schema/tx"
        xmlns:context="http://www.springframework.org/schema/context"
        xsi:schemaLocation="
     http://www.springframework.org/schema/beans
     http://www.springframework.org/schema/beans/spring-beans-3.0.xsd
     http://www.springframework.org/schema/tx
     http://www.springframework.org/schema/tx/spring-tx-3.0.xsd
     http://www.springframework.org/schema/context
     http://www.springframework.org/schema/context/ spring-context-
3.0.xsd">

    <context:property-placeholder location="classpath:*.properties"/>
    <!--设置 redis 连接池-->
    <bean id="poolConfig"  class="redis.clients.jedis.JedisPoolConfig">
        <property name="maxIdle" value="${redis.maxIdle}"></property>
        <property name="minIdle" value="${redis.minIdle}"></property>
        <property name="maxTotal" value="${redis.maxTotal}"></property>
        <property name="maxWaitMillis"
value="${redis.maxWaitMillis}"></property>
        <property name="testOnBorrow"
value="${redis.testOnBorrow}"></property>
    </bean>
    <!--链接 redis-->
    <bean id="connectionFactory"
```

```
    class="org.springframework.data.redis.connection.jedis.
JedisConnectionFactory">
            <property name="hostName" value="${redis.host}"></property>
            <property name="port" value="${redis.port}"></property>
            <property name="password" value="${redis.password}"></property>
            <property name="poolConfig" ref="poolConfig"></property>
        </bean>
        <!-- redis 工具类 -->
        <bean id="redisTemplate"
class="org.springframework.data.redis.core.RedisTemplate"
     p:connection-factory-ref="connectionFactory" >
        </bean>
    </beans>
```

在 src\main\resources\applicationContext.xml 配置文件中导入 spring-redis.xml 配置，具体代码如下：

```
    <import resource="spring-redis.xml"/>
```

所有的配置文件添加完成后，在 src\main\test\com\ay\test 目录下创建测试类 RedisTest.java，具体代码如下：

```
    /** redis 缓存测试
     * @author Ay
     * @create 2018/07/08
     **/
    public class RedisTest extends BaseJunit4Test{
        @Resource
        private RedisTemplate redisTemplate;

        @Test
        public void testRedis(){
            redisTemplate.opsForValue().set("name", "ay");
              String name = (String) redisTemplate.opsForValue().get("name");
            System.out.println("value of name is:" + name);
        }
    }
```

RedisTemplate 和 StringRedisTemplate 都是 Spring Data Redis 为我们提供的两个模板类用来对数据进行操作，其中 StringRedisTemplate 只针对键值都是字符串的数据进行操作。在应用启动的时候，Spring 会初始化这两个模板类，通过@Resource 注解注入即可使用。

RedisTemplate 和 StringRedisTemplate 除了提供 opsForValue 方法用来操作简单属性数据之外，还提供了以下其他主要数据的访问方法：

（1）opsForList：操作含有 list 的数据。
（2）opsForSet：操作含有 set 的数据。

（3）opsForZSet：操作含有 zset（有序 set）的数据。

（4）opsForHash：操作含有 hash 的数据。

当数据存放到 Redis 的时候，键（key）和值（value）都是通过 Spring 提供的 Serializer 序列化到数据库的。RedisTemplate 默认使用 JdkSerializationRedisSerializer，而 StringRedisTemplate 默认使用 StringRedisSerializer。

RedisTest 开发完成后，运行测试类 RedisTest 的 testRedis()方法，找到 Redis 客户端程序 redis-cli.exe，当在客户端输入命令：“get name”后可以获取到值，同时可以在 IDEA 控制台中看到打印的信息，表示 SSM 框架成功地集成了 Redis 缓存。具体如图 11-16 和图 11-17 所示。

```
127.0.0.1:6379> get name
"a1"
127.0.0.1:6379>
```

图 11-16　SSM 框架集成 Redis 测试

```
11:23:33,091 DEBUG RedisConnectionUtils:205 - Closing Red
11:23:33,092 DEBUG RedisConnectionUtils:125 - Opening Red
11:23:33,096 DEBUG RedisConnectionUtils:205 - Closing Red
value of name is:ay
11:23:33,099 DEBUG AbstractDirtiesContextTestExecutionLis
11:23:33,101 DEBUG AbstractDirtiesContextTestExecutionLis
11:23:33,103  INFO GenericApplicationContext:989 - Closin
```

图 11-17　控制台打印信息

11.3.4　设计 Redis 数据结构

高并发点赞项目的 Redis 数据结构采用多个 set 类型的集合来存放，具体如图 11-18 所示。

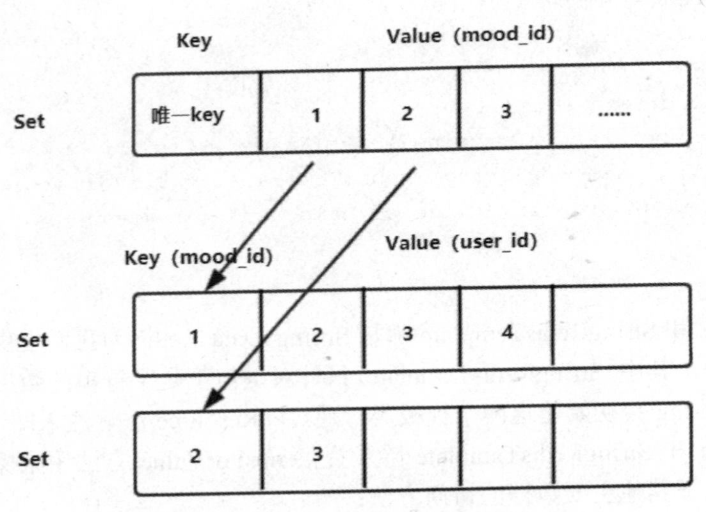

图 11-18　Redis 数据结构设计

首先，用一个 set 集合来存放所有被点赞的说说 id，key 可以是自己约定的全局唯一的

key 即可，而 value 为所有被点赞的说说 id（mood_id）。然后用 n 个 set 集合来存放每条说说被哪些用户点赞的记录。如图 11-18 所示，mood_id = 1 被用户 id = {2，3，4}点赞，mood_id = 2 被用户 id = {3}点赞。如果需要获取某个 mood_id 被点赞的次数，只需要统计 set 集合的 size()即可。通过上面的 Redis 数据结构设计，基本可以满足我们的需求。

11.3.5　代码实现

在 src\main\java\com\ay\service\MoodService.java 文件中添加如下代码：

```
/**
 * 描述：说说接口
 * @author Ay
 * @date 2018/1/6
 */
public interface MoodService {
    //省略代码
    boolean praiseMoodForRedis(String userId, String moodId);
    List<MoodDTO> findAllForRedis();
}
```

在 MoodService 接口中添加两个方法：praiseMoodForRedis()和 findAllForRedis()。praiseMoodForRedis()用来把说说点赞的记录保存到 Redis 缓存中，findAllForRedis 用来查询所有的说说列表，包括从缓存中查询说说被点赞的次数。

在 src\main\java\com\ay\service\impl\MoodServiveImpl.java 文件中添加如下代码：

```
/**
 * 描述：说说服务类
 * @author Ay
 * @date 2018/1/6
 */
@Service
public class MoodServiveImpl implements MoodService {
    //省略代码
    @Resource
    private RedisTemplate redisTemplate;
    //key 命名规范：项目名称 + 模块名称 + 具体内容
    private static final String PRAISE_HASH_KEY =
"springmv.mybatis.boot.mood.id.list.key";

    public boolean praiseMoodForRedis(String userId, String moodId) {
        //1. 存放到 set 集合中
        redisTemplate.opsForSet().add(PRAISE_HASH_KEY , moodId);
        //2. 存放到 set 中
        redisTemplate.opsForSet().add(moodId,userId);
        return false;
```

```
    }

    @Resource
    private UserService userService;

    public List<MoodDTO> findAllForRedis() {
        List<Mood> moodList = moodDao.findAll();
        if(CollectionUtils.isEmpty(moodList)){
            return Collections.EMPTY_LIST;
        }
        List<MoodDTO> moodDTOList = new ArrayList<MoodDTO>();
        for(Mood mood: moodList){
            MoodDTO moodDTO = new MoodDTO();
            moodDTO.setId(mood.getId());
            moodDTO.setUserId(mood.getUserId());
            //right = 总点赞数量：数据库的点赞数量 + redis 的点赞数量
            moodDTO.setPraiseNum(mood.getPraiseNum() +
redisTemplate.opsForSet().size(mood.getId()).intValue());
            moodDTO.setPublishTime(mood.getPublishTime());
            moodDTO.setContent(mood.getContent());
            //通过 userID 查询用户
            User user =  userService.find(mood.getUserId());
            //用户名
            moodDTO.setUserName(user.getName());
            //账户
            moodDTO.setUserAccount(user.getAccount());
            moodDTOList.add(moodDTO);
        }
        return moodDTOList;
    }
}
```

MoodServiveImpl 类主要实现 MoodServive 接口中的 praiseMoodForRedis() 和 findAllForRedis()方法。在 praiseMoodForRedis()方法中，处理逻辑比较简单：

（1）保存 mood_id 到 set 集合中。

（2）保存 mood_id 和点赞的 user_id 到 set 集合中。

注意，这里的 set 集合不是同一个，而是分开存储的。有多少条说说被点赞，在 Redis 缓存中就存在多少个 set 集合，每条说说的点赞记录分别用一个 set 集合存储，这样可以保证每个 set 集合所占空间不会过大，同时在查询和统计的时候，处理速度也会比较快。

在 src\main\java\com\ay\controller\MoodController.java 文件中添加如下代码：

```
/**
 * 描述：说说控制层
 * @author Ay
```

```
 * @date 2018/1/6
 */
@Controller
@RequestMapping("/mood")
public class MoodController {

    @Resource
    private MoodService moodService;

        //省略代码

    @GetMapping(value = "/{moodId}/praiseForRedis")
    public String praiseForRedis(Model model, @PathVariable
(value="moodId")String moodId,
                                 @RequestParam(value="userId")String userId){
        //方便使用，随机生成用户 id
        Random random = new Random();
        userId = random.nextInt(100) + "";

        boolean isPraise = moodService.praiseMoodForRedis(userId, moodId);
        //查询所有的说说数据
        List<MoodDTO> moodDTOList = moodService.findAllForRedis();
        model.addAttribute("moods",moodDTOList);
        model.addAttribute("isPraise", isPraise);
        return "mood";
    }
}
```

MoodController 主要是控制层的代码，用来接收前端的点赞请求。如下代码主要是为了随机生成 user_id，然后给某一条说说点赞。纯粹是为了简化逻辑而开发的，在真实的项目中并不是这样的逻辑。

```
//方便使用，随机生成用户 id
Random random = new Random();
userId = random.nextInt(100) + "";
```

在 src\main\webapp\WEB-INF\views\mood.jsp 文件中添加如下代码：

```
<%@page language="java" contentType="text/html; charset=UTF-8"
      pageEncoding="UTF-8" isELIgnored="false"%>
<%@ taglib uri="http://java.sun.com/jsp/jstl/core" prefix="c"%>
<%@ taglib prefix="fmt" uri="http://java.sun.com/jstl/fmt" %>
<!DOCTYPE HTML>
<html>
<head>
    <title>Getting Started: Serving Web Content</title>
    <meta http-equiv="Content-Type" content="text/html; charset=UTF-8" />
```

```
</head>
<body>

<div id="moods">
    <b>说说列表:</b><br>
    <c:forEach items="${moods}" var="mood">
        ------------------------------------
        <br>
        <b>用户：</b><span id="account">${mood.userName}</span><br>
        <b>说说内容：</b><span id="content">${mood.content}</span><br>
        <b>发表时间：</b>
        <span id="publish_time">
              ${mood.publishTime}
        </span><br>
        <b>点赞数：</b><span id="praise_num">${mood.praiseNum}</span><br>
        <div style="margin-left: 350px">
            <%--  传统点赞请求 -->
            <%--<a id="praise" href="/mood/${mood.id}/praise?userId=
${mood.userId}">赞</a>--%>
          <%--  引入 redis 缓存的点赞请求 -->
            <a id="praise" href="/mood/${mood.id}/praiseForRedis?userId=
${mood.userId}">赞</a>
        </div>
    </c:forEach>
</div>
</body>
<script></script>
</html>
```

11.3.6 集成 Quartz 定时器

Quartz 是一个完全由 Java 编写的开源任务调度的框架，通过触发器设置作业定时运行规则，控制作业的运行时间。Quartz 定时器作用很多，比如，定时发送信息和定时生成报表等。

Quartz 框架主要核心组件包括调度器、触发器和作业。调度器作为作业的总指挥，触发器作为作业的操作者，作业为应用的功能模块。其关系如图 11-19 所示。

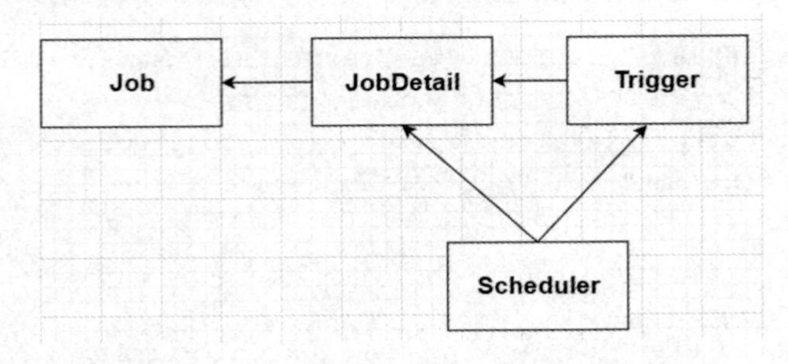

图 11-19 Quartz 各个组件的关系

Job 是一个接口，该接口只有一个方法 execute，被调度的作业（类）需实现该接口中 execute()方法，JobExecutionContext 类提供了调度上下文的各种信息。每次执行该 Job 均重新创建一个 Job 实例。Job 的源码如下：

```
public interface Job {
    void execute(JobExecutionContext var1) throws JobExecutionException;
}
```

Quartz 在每次执行 Job 时，都重新创建一个 Job 实例，所以它不直接接收一个 Job 的实例，相反它接收一个 Job 实现类，以便运行时通过 newInstance()的反射机制实例化 Job。因此需要通过一个类来描述 Job 的实现类及其他相关的静态信息，如 Job 名字、描述、关联监听器等信息，JobDetail 承担了这一角色。JobDetail 用来保存作业的详细信息，一个 JobDetail 可以有多个 Trigger，但是一个 Trigger 只能对应一个 JobDetail

Trigger 触发器描述触发 Job 的执行规则，主要有 SimpleTrigger 和 CronTrigger 这两个子类。当仅需触发一次或者以固定时间间隔周期执行时，SimpleTrigger 是最适合的选择；而 CronTrigger 则可以通过 Cron 表达式定义出各种复杂时间规则的调度方案：如每早晨 9:00 执行，周一、周三、周五下午 5:00 执行等。Cron 表达式定义如下：

```
CronTrigger 配置格式：
格式：[秒] [分] [小时] [日] [月] [周] [年]
0 0 12 * * ?                    每天 12 点触发
0 15 10 ? * *                   每天 10 点 15 分触发
0 15 10 * * ?                   每天 10 点 15 分触发
0 15 10 * * ? *                 每天 10 点 15 分触发
0 15 10 * * ? 2005              2005 年每天 10 点 15 分触发
0 * 14 * * ?                    每天下午的 2 点到 2 点 59 分每分钟触发
0 0/5 14 * * ?                  每天下午的 2 点到 2 点 59 分(整点开始，每隔 5 分钟触发)
0 0/5 14,18 * * ?               在每天下午 2 点到 2:55 期间和下午 6 点到 6:55 期间的每 5
                                  分钟触发
0 0-5 14 * * ?                  每天下午的 2 点到 2 点 05 分每分钟触发
0 10,44 14 ? 3 WED              3 月份每周三下午的 2 点 10 分和 2 点 44 分触发
0 15 10 ? * MON-FRI             从周一到周五每天上午的 10 点 15 分触发
0 15 10 15 * ?                  每月 15 号上午 10 点 15 分触发
0 15 10 L * ?                   每月最后一天的 10 点 15 分触发
0 15 10 ? * 6L                  每月最后一周的星期五的 10 点 15 分触发
0 15 10 ? * 6L 2002-2005        从 2002 年到 2005 年每月最后一周的星期五的 10 点 15 分触发
0 15 10 ? * 6#3                 每月的第三周的星期五开始触发
0 0 12 1/5 * ?                  每月的第一个中午开始每隔 5 天触发一次
0 11 11 11 11 ?                 每年的 11 月 11 号 11 点 11 分触发(光棍节)
```

Scheduler 负责管理 Quartz 的运行环境，Quartz 是基于多线程架构的，它启动的时候会初始化一套线程，这套线程会用来执行一些预置的作业。Trigger 和 JobDetail 可以注册到 Scheduler 中，Scheduler 可以将 Trigger 绑定到某一 JobDetail 中，这样当 Trigger 触发时，对应的 Job 就被执行。Scheduler 拥有一个 SchedulerContext，它类似于 ServletContext，保存着

Scheduler 上下文信息，Job 和 Trigger 都可以访问 SchedulerContext 内的信息。Scheduler 使用一个线程池作为任务运行的基础设施，任务通过共享线程池中的线程提高运行效率。

了解完 Quartz 定时器的基本原理后，在 src\main\java\com\ay\job 目录下创建定时器类 PraiseDataSaveDBJob.java，具体代码如下：

```
    /**
     * 描述：定时器
     * @author Ay
     * @date 2018/1/6
     */
    @Component
    @Configurable
    @EnableScheduling
    public class PraiseDataSaveDBJob {
        @Resource
        private RedisTemplate redisTemplate;
        private static final String PRAISE_HASH_KEY =
"springmv.mybatis.boot.mood.id.list.key";
        @Resource
        private UserMoodPraiseRelService userMoodPraiseRelService;
        @Resource
        private MoodService moodService;
        //每 10 秒执行一次，真实项目当中，可以把定时器的执行计划时间设置长一点
        //比如说每天晚上凌晨 2 点跑一次.
        @Scheduled(cron = "*/10 *  *  * * * ")
        public void savePraiseDataToDB2(){
            //step1：在 redis 缓存中所有所有被点赞的说说 id
            Set<String> moods = redisTemplate.opsForSet().members
(PRAISE_HASH_KEY);
            if(CollectionUtils.isEmpty(moods)){
                return;
            }
            for(String moodId: moods){
                if(redisTemplate.opsForSet().members(moodId) == null){
                    continue;
                }else {
                    //step2：从 Redis 缓存中，通过说说 id 获取所有点赞的用户 id 列表
                    Set<String> userIds = redisTemplate.opsForSet().members
(moodId);
                    if(CollectionUtils.isEmpty(userIds)){
                        continue;
                    }else{
                        //step3：循环保存 mood_id 和 user_id 的关联关系到 MySQL 数据库
                        for(String userId:userIds){
```

```
                UserMoodPraiseRel userMoodPraiseRel = new
UserMoodPraiseRel();
                userMoodPraiseRel.setMoodId(moodId);
                userMoodPraiseRel.setUserId(userId);
                //保存说说与用户的关联关系
                userMoodPraiseRelService.save(userMoodPraiseRel);
            }
            Mood mood = moodService.findById(moodId);
            //step4: 更新说说点赞数量
            //说说的总点赞数量 = Redis 点赞数量 + 数据库的点赞数量
            mood.setPraiseNum(mood.getPraiseNum() +
redisTemplate.opsForSet().size(moodId).intValue());
            moodService.update(mood);
            //step5: 清除 Redis 缓存中的数据
            redisTemplate.delete(moodId);
        }
      }
    }
    //step6: 清除 Redis 缓存中的数据
    redisTemplate.delete(PRAISE_HASH_KEY);

  }
}
```

- @Configurable：加上此注解的类相当于XML配置文件可以被Spring扫描初始化。
- @EnableScheduling：通过在配置类注解@EnableScheduling来开启对计划任务的支持，然后在要执行计划任务的方法上注解@Scheduled，声明这是一个计划任务。
- @Scheduled：注解为定时任务，Cron表达式里写执行的时机。

使用 Quartz 定时器有两种方式：一是 XML 配置；二是注解方式。本书使用注解的方式开发代码。在定时器类 PraiseDataSaveDBJob 中定义了 savePraiseDataToDB2()方法，该方法通过@Scheduled 注解定义了方法的执行计划——每 10 秒执行一次，当然这样的配置只是方便测试，在真实项目当中，可以把定时器的执行计划时间设置晚一点，比如每天晚上凌晨 2 点开始执行。savePraiseDataToDB2()方法的逻辑相对比较简单，具体说明如下：

（1）从 Redis 缓存中获取所有被点赞的 mood_id。
（2）通过 mood_id 从 Redis 缓存中获取所有点赞的 user_id 列表。
（3）循环保存 mood_id 和 user_id 的关联关系到 MySQL 数据库。
（4）更新说说的点赞数量。
（5）清除 Redis 缓存中的数据。

11.3.7 测试

所有的代码开发完成之后，重新启动项目 springmvc-mybatis-book，在浏览器中输入访问 URL：http://localhost:8080/mood/findAll，查询所有的说说列表，不断地单击第一条说说的赞功能，便可以看到第一条说说的点赞数量不断增加，同时查看 MySQL 数据库表 mood 和 user_mood_praise_rel，可以看到 mood 表的 praise_num 的点赞数量不断被更新，而 user_mood_praise_rel 表中 mood_id 和 user_id 的关联关系数据每隔 10s 也会被保存到表中。具体如图 11-20 和图 11-21 所示。

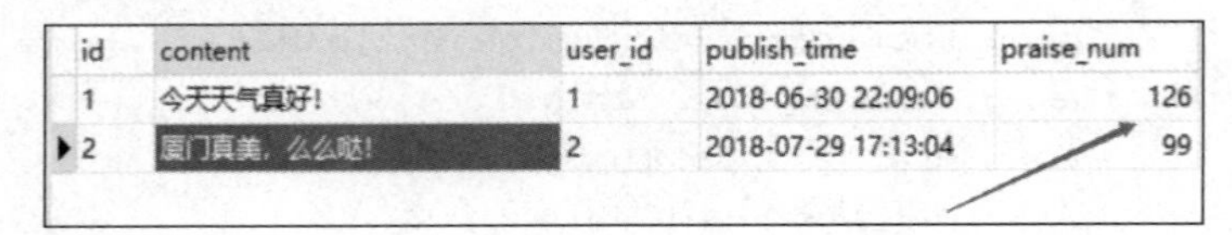

id	content	user_id	publish_time	praise_num
1	今天天气真好!	1	2018-06-30 22:09:06	126
2	厦门真美，么么哒!	2	2018-07-29 17:13:04	99

图 11-20　mood 表的 praise_num 的点赞数量不断被更新

id	user_id	mood_id
10	1	1
28	36	1
29	43	1
30	38	1
31	66	1
32	54	1
33	82	1
34	42	1
35	27	1

图 11-21　mood_id 和 user_id 的关联关系数据

11.4 集成 ActiveMQ

11.4.1 概述

前文，已经总结了传统的点赞功能实现所暴露的问题，主要有以下几点：

（1）高并发请求下，服务器频繁创建线程。

（2）高并发请求下，数据库连接池中的连接数有限。

（3）高并发请求下，点赞功能是同步处理等。

我们通过引入 Redis 缓存避免高并发写数据库而造成数据库压力，同时引入 Redis 缓存提高读的性能，基本可以解决问题（2），本节我们主要针对问题（3）提供解决方案，具体如图 11-22 所示。

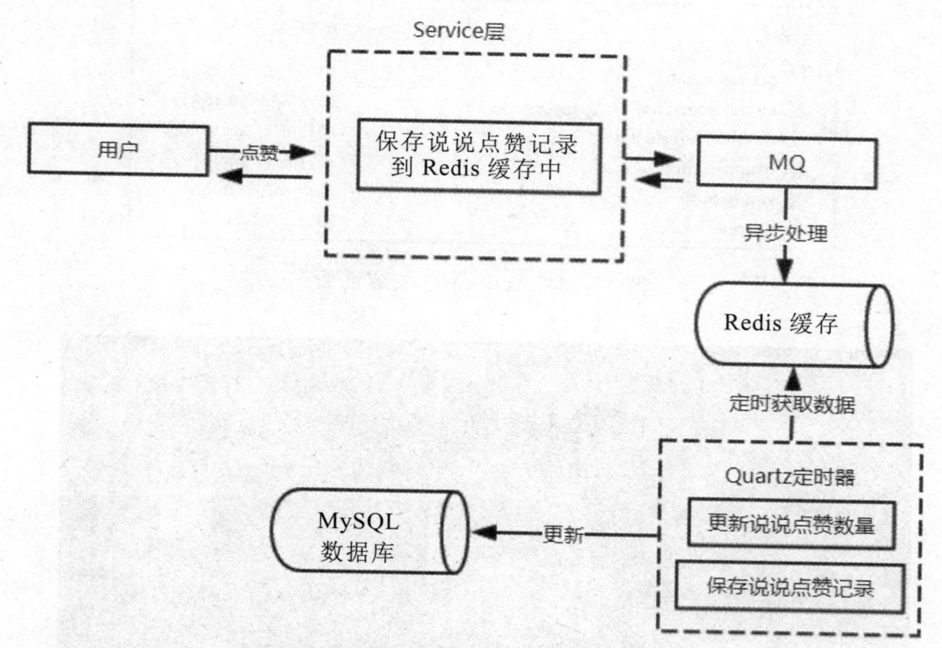

图 11-22 高并发点赞项目解决方案

为了解决高并发请求下点赞功能同步处理所带来的服务器压力（Redis 缓存的压力或数据库压力等），我们引入 MQ 消息中间件进行异步处理，用户每次点赞都会推送消息到 MQ 服务器并及时返回，这样用户的点赞请求就及时结束，避免了点赞请求线程占用时间长的问题。与此同时，MQ 消息中间件接收到消息后，会按照“自己的方式”及时消费，还可以用 MQ 消息中间件来限制流量并进行异步处理等。

11.4.2 ActiveMQ 的安装

MQ 英文名是 MessageQueue，中文名是消息队列，是一个消息的接受和转发的容器，可用于消息推送。ActiveMQ 是 Apache 提供的一个开源的消息系统，完全采用 Java 来实现，因此，它能很好地支持 J2EE 提出的 JMS（Java Message Service，即 Java 消息服务）规范。

安装 ActiveMQ 之前，我们需要到官网（http://activemq.apache.org/activemq-5150-release.html）下载，本书使用 apache-activemq-5.15.0 这个版本进行讲解。ActiveMQ 具体安装步骤如下：

步骤 01 将官网下载的安装包 apache-activemq-5.15.0-bin.zip 解压。

步骤 02 打开解压的文件夹，进入 bin 目录，根据计算机操作系统是 32 位还是 64 位，选择进入【win32】文件夹或者【win64】文件夹。

步骤 03 双击【activemq.bat】，即可启动 ActiveMQ，如图 11-23 所示。当看到如图 11-24 所示的启动信息时，代表 ActiveMQ 安装成功。从图中可以看出，ActiveMQ 默认启动到 8161 端口。

名称	修改日期
activemq.bat	2017/6/27 13:48
InstallService.bat	2017/6/27 13:48
UninstallService.bat	2017/6/27 13:48
wrapper.conf	2017/6/27 13:48
wrapper.dll	2017/6/27 13:43
wrapper.exe	2017/6/27 13:43

图 11-23　ActiveMQ 安装界面

```
ActiveMQ
jvm 1    | INFO | Apache ActiveMQ 5.15.0 (localhost, ID:DESKTOP-9RV1IBJ-51478-1511882105912-0:1) is starting
jvm 1    | INFO | Listening for connections at: tcp://DESKTOP-9RV1IBJ:61616?maximumConnections=1000&wireFormat.m
axFrameSize=104857600
jvm 1    | INFO | Connector openwire started
jvm 1    | INFO | Listening for connections at: amqp://DESKTOP-9RV1IBJ:5672?maximumConnections=1000&wireFormat.m
axFrameSize=104857600
jvm 1    | INFO | Connector amqp started
jvm 1    | INFO | Listening for connections at: stomp://DESKTOP-9RV1IBJ:61613?maximumConnections=1000&wireFormat
.maxFrameSize=104857600
jvm 1    | INFO | Connector stomp started
jvm 1    | INFO | Listening for connections at: mqtt://DESKTOP-9RV1IBJ:1883?maximumConnections=1000&wireFormat.m
axFrameSize=104857600
jvm 1    | INFO | Connector mqtt started
jvm 1    | WARN | ServletContext@o.e.j.s.ServletContextHandler@4c14b62b{/,null,STARTING} has uncovered http meth
ods for path: /
jvm 1    | INFO | Listening for connections at ws://DESKTOP-9RV1IBJ:61614?maximumConnections=1000&wireFormat.max
FrameSize=104857600
jvm 1    | INFO | Connector ws started
jvm 1    | INFO | Apache ActiveMQ 5.15.0 (localhost, ID:DESKTOP-9RV1IBJ-51478-1511882105912-0:1) started
jvm 1    | INFO | For help or more information please see: http://activemq.apache.org
jvm 1    | INFO | No Spring WebApplicationInitializer types detected on classpath
jvm 1    | INFO | ActiveMQ WebConsole available at http://0.0.0.0:8161/
jvm 1    | INFO | ActiveMQ Jolokia REST API available at http://0.0.0.0:8161/api/jolokia/
jvm 1    | INFO | Initializing Spring FrameworkServlet 'dispatcher'
jvm 1    | INFO | No Spring WebApplicationInitializer types detected on classpath
jvm 1    | INFO | jolokia-agent: Using policy access restrictor classpath:/jolokia-access.xml
搜狗拼音输入法 全：
```

图 11-24　ActiveMQ 启动成功界面

步骤 04 安装成功之后，在浏览器中输入 http://localhost:8161/admin 链接访问，第一次访问需要输入用户名 admin 和密码 admin 进行登录，登录成功之后，就可以看到 ActiveMQ 的首页，具体如图 11-25 所示。

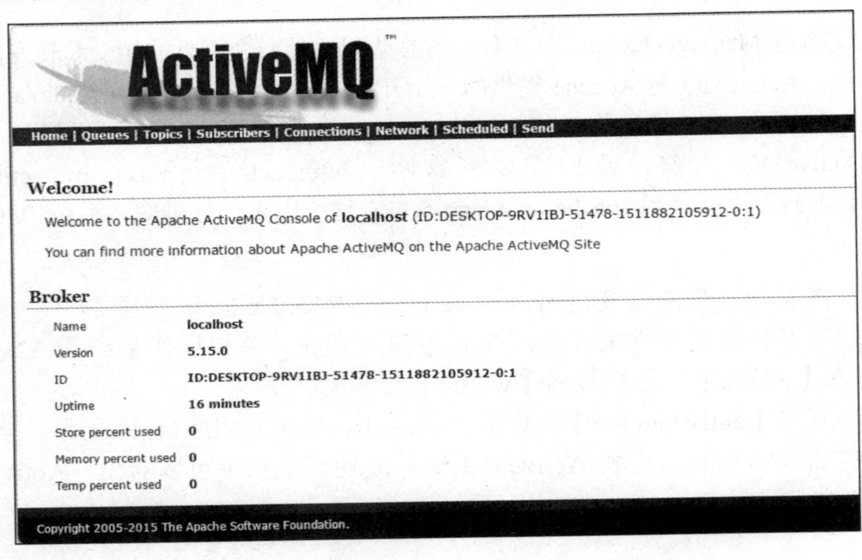

图 11-25　ActiveMQ 首页

11.4.3 集成 ActiveMQ

在 SSM 框架中集成 ActiveMQ 缓存，首先需要在 pom.xml 文件中引入所需的依赖，具体代码如下：

```
<!-- active mq start -->
<dependency>
     <groupId>org.springframework</groupId>
     <artifactId>spring-jms</artifactId>
     <version>${spring.version}</version>
</dependency>
<dependency>
     <groupId>org.apache.activemq</groupId>
     <artifactId>activemq-all</artifactId>
     <version>5.11.2</version>
     <exclusions>
       <exclusion>
         <artifactId>spring-context</artifactId>
         <groupId>org.springframework</groupId>
       </exclusion>
       <exclusion>
         <groupId>org.apache.geronimo.specs</groupId>
         <artifactId>geronimo-jms_1.1_spec</artifactId>
       </exclusion>
     </exclusions>
</dependency>
<dependency>
     <groupId>javax.jms</groupId>
     <artifactId>javax.jms-api</artifactId>
     <version>2.0.1</version>
</dependency>
<!-- active mq end -->
```

依赖添加完成之后，需要在\src\main\resources 目录下创建 ActiveMQ 配置文件 spring-jms.xml，具体代码如下：

```
<?xml version="1.0" encoding="UTF-8"?>
<beans xmlns:xsi="http://www.w3.org/2001/XMLSchema-instance"
       xmlns="http://www.springframework.org/schema/beans"
xmlns:context="http://www.springframework.org/schema/context"
       xmlns:jms="http://www.springframework.org/schema/jms"
       xsi:schemaLocation="http://www.springframework.org/schema/beans
          http://www.springframework.org/schema/beans/spring-beans-4.0.xsd
          http://www.springframework.org/schema/context
          http://www.springframework.org/schema/context/ spring-context-
4.0.xsd
```

```
            http://www.springframework.org/schema/jms
            http://www.springframework.org/schema/jms/ spring-jms-4.0.xsd">
    <bean id="connectionFactory"
          class="org.springframework.jms.connection.
CachingConnectionFactory">
        <description>JMS 连接工厂</description>
        <property name="targetConnectionFactory">
            <bean class="org.apache.activemq.spring.
ActiveMQConnectionFactory">
                <property name="brokerURL" value="${activemq_url} " />
                <property name="userName" value="${activemq_username}" />
                <property name="password" value="${activemq_password}" />
            </bean>
        </property>
        <property name="sessionCacheSize" value="100" />
    </bean>

    <!-- Spring JmsTemplate 的消息生产者 start-->
    <bean id="jmsTemplate" class=
"org.springframework.jms.core.JmsTemplate">
        <description>队列模式模型</description>
        <constructor-arg ref="connectionFactory" />
        <property name="receiveTimeout" value="10000" />
        <!-- 如果为 True，则是 Topic；如果是 false 或者默认，则是 queue -->
        <property name="pubSubDomain" value="false" />
    </bean>

    <!-- 消息消费者 start-->
    <!-- 定义 Queue 监听器 -->
    <jms:listener-container destination-type="queue"
                        container-type="default" connection-factory=
"connectionFactory"
                        acknowledge="auto">
    <!-- 可写多个监听器 -->
    <jms:listener destination="ay.queue.high.concurrency.praise" ref=
"moodConsumer" />
    </jms:listener-container>
    <!-- 消息消费者 end -->
</beans>
```

在配置文件 spring-jms.xml 中，首先定义了 ActiveMQ 的连接信息，然后定义了 JmsTemplate 工具类，该工具类是 Spring 框架提供的，利用 JmsTemplate 可以很方便地发送和接收消息。最后，我们定义消费者类 moodConsumer，同时消费者监听的是 ay.queue.high.concurrency.praise 这个 topic，当有生产者 Producer 往该队列推送消息时，消费者 Consumer 就可以监听到该消息，并做相应的逻辑处理。

在\src\main\resources 目录下创建配置文件 activemq.properties，具体代码如下：

```
### active mq 服务器地址
activemq_url=tcp://localhost:61616
### 服务器用户名
activemq_username=admin
### 服务器密码
activemq_password=admin
```

activemq.properties 属性文件主要配置 ActiveMQ 的连接信息，供 spring-jms.xml 配置文件使用。

spring-jms.xml 开发完成之后，在 applicationContext.xml 配置文件中使用<import>标签引入入 spring-jms.xml 配置文件，具体代码如下：

```
<import resource="spring-jms.xml"/>
```

11.4.4　ActiveMQ 异步消费

上一节，已经在项目中集成了 ActiveMQ 消息中间件，同时开发了相关的配置文件，这一节主要利用 ActiveMQ 实现点赞功能的异步处理。具体如下：

首先，在 src\main\java\com\ay\mq 目录下创建生产者类 MoodProducer，具体代码如下：

```
/**
 * 生产者 jmsTemplate
 * @author Ay
 * @date 2017/11/30
 */
@Component
public class MoodProducer {

    @Resource
    private JmsTemplate jmsTemplate;

private Logger log = Logger.getLogger(this.getClass());

    public void sendMessage(Destination destination, final MoodDTO mood) {
        //记录日志
        log.info("生产者--->>>用户 id: " + mood.getUserId() + " 给说说 id: " +
mood.getId() + " 点赞");
        //mood 实体需要实现 Serializable 序列化接口
        jmsTemplate.convertAndSend(destination, mood);
    }
}
```

MoodProducer 类提供 sendMessage 方法用来发送消息，第一个参数是 destination，主要用来指定队列的名称，第二个参数是 mood 说说实体。通过调用 jmsTemplate 工具类的

convertAndSend 方法发送消息。需要注意的是，MoodDTO 说说实体需要实现序列化接口 Serializable，具体代码如下：

```
/**
 * 描述：说说
 * Created by Ay on 2017/9/16
 */
public class MoodDTO implements Serializable{
   //省略代码
}
```

其次，在 src\main\java\com\ay\mq 目录下创建消费者类 MoodConsumer。

```
/**
 * 消费者
 * @author Ay
 * @date  2017/11/30
 */
@Component
public class MoodConsumer implements MessageListener {

   private static final String PRAISE_HASH_KEY =
 "springmv.mybatis.boot.mood.id.list.key";

   private Logger log = Logger.getLogger(this.getClass());

   @Resource
   private RedisTemplate redisTemplate;

   public void onMessage(Message message) {
      try{
         //从 message 对象中获取说说实体
         MoodDTO moodDTO = (MoodDTO)((ActiveMQObjectMessage)
message).getObject();
         //1. 存放到 set 中
         redisTemplate.opsForSet().add(PRAISE_HASH_KEY , moodDTO.getId());
         //2. 存放到 set 中
         redisTemplate.opsForSet().add(moodDTO.getId(),
moodDTO.getUserId());
         log.info("消费者--->>>用户 id: " + mood.getUserId() + " 给说说 id:" +
mood.getId() + " 点赞");
      }catch (Exception e){
         System.out.println(e);
      }
   }
}
```

消费者类 MoodConsumer 实现 MessageListener 接口，完成对消息的监听和接收。消息有两种接收方式：同步接收和异步接收。

- **同步接收**：主线程阻塞式等待下一个消息的到来，可以设置timeout，超时则返回null。
- **异步接收**：主线程设置MessageListener，然后继续做自己的事，子线程负责监听。

同步接收又称为阻塞式接收，异步接收又称为事件驱动接收。同步接收，是在获取 MessageConsumer 实例之后调用以下的 API：

- receive()：获取下一个消息。该调用将导致无限期的阻塞，直到有新的消息产生。
- receive(long timeout)：获取下一个消息。该调用可能导致一段时间的阻塞，直到超时或者有新的消息产生。超时则返回null。
- receiveNoWait()：获取下一个消息。这个调用不会导致阻塞，如果没有下一个消息，直接返回null。

异步接收是在获取 MessageConsumer 实例之后调用下面的 API：

- setMessageListener(MessageListener)：设置消息监听器。MessageListener是一个接口，只定义了一个方法，即onMessage(Message message)回调方法，当有新的消息产生时，该方法会被自动调用。

可见，为实现异步接收只需要对 MessageListener 进行实现，然后设置为 Consumer 实例的 messageListener 即可。

最后，修改 MoodServiveImpl 类中的 praiseMoodForRedis()方法，将其改成异步处理方式，具体代码如下：

```
/**
 * 描述：说说服务类
 * @author Ay
 * @date 2018/1/6
 */
@Service
public class MoodServiveImpl implements MoodService {

    @Resource
    private MoodProducer moodProducer;
       @Resource
    private RedisTemplate redisTemplate;

    //队列
    private static Destination destination =
new ActiveMQQueue("ay.queue.high.concurrency.praise");
```

```
    public boolean praiseMoodForRedis(String userId, String moodId) {
        //修改为异步处理方式
        MoodDTO moodDTO = new MoodDTO();
        moodDTO.setUserId(userId);
        moodDTO.setId(moodId);
        //发送消息
        moodProducer.sendMessage(destination, moodDTO);

//      //1. 存放到 hashset 中
//      redisTemplate.opsForSet().add(PRAISE_HASH_KEY , moodId);
//      //2. 存放到 set 中
//      redisTemplate.opsForSet().add(moodId,userId);
        return false;
    }
}
```

11.4.5 测试

代码开发完成之后，重新启动项目 springmvc-mybatis-book，在浏览器中输入请求 URL: http://localhost:8080/mood/findAll，给某条说说点赞，如果在控制台看到如图 11-26 和图 11-27 所示的信息，代表使用 MQ 异步消费开发成功。

```
DEBUG DispatcherServlet:979 - Last-Modified value for [/mood/1
:1531660218776
 INFO MoodProducer:26 - 生产者--->>>用户id: 67 给说说id: 1 点赞
DEBUG JmsTemplate:502 - Executing callback on JMS Session: Ca
DEBUG JmsTemplate:606 - Sending created message: ActiveMQObje
DEBUG DefaultMessageListenerContainer:306 - Received message
```

图 11-26 生产者打印信息

```
DEBUG RedisConnectionUtils:125 - Opening RedisConnection
DEBUG RedisConnectionUtils:[illegible] - Closing Redis Connection
 INFO MoodConsumer:36 - 消费者--->>>用户id: 77 给说说id: 1 点赞
DEBUG SqlSessionUtils:97 - Creating a new SqlSession
DEBUG SqlSessionUtils:148 - SqlSession [org.apache.ibatis.ses
DEBUG DataSourceUtils:114 - Fetching JDBC Connection from Dat
```

图 11-27 消费者打印信息

11.5 思考与练习

1. 简述 Redis 缓存的作用。
2. 动手在 Windows 操作系统上安装 Redis 缓存。
3. 简述 Redis 五种数据结构。

4. 简述 Quartz 定时器及其作用。
5. 简述 Quartz 各个组件的关系。
6. 简述 ActiveMQ 消息中间件的作用。
7. 动手安装 ActiveMQ 消息中间件。
8. 动手在 MYSQL 数据库中创建用户表、说说表等。
9. 思考如何提高数据库查询性能。

参考文献

[1] https://baike.baidu.com/item/Intellij%20IDEA/9548353?fr=aladdin

[2] https://baike.baidu.com/item/tomcat/255751?fr=aladdin

[3] https://baike.baidu.com/item/Maven/6094909?fr=aladdin

[4] https://projects.spring.io/spring-framework/

[5] 汪云飞. Java EE 开发的颠覆者：Spring Boot 实战[M]. 北京：电子工业出版社，2016.

[6] 郝佳. Spring 源码深度解析[M]. 北京：人民邮电出版社，2013.

[7] 王富强. Spring Boot 揭秘：快速构建微服务体系[M]. 北京：机械工业出版社，2016.

[8] https://spring.io/guides

[9] https://baike.baidu.com/item/ApplicationContext/1129418

[10] https://baike.baidu.com/item/DispatcherServlet/12740507?fr=aladdin

[11] 刘伟. 设计模式[M]. 北京：清华大学出版社，2011.

[12] 徐郡明. MyBatis 技术内幕[M]. 北京：电子工业出版社，2017.

[13] 郝佳. Spring 源码深度解析[M]. 北京：人民邮电出版社，2013.

[14] http://www.mybatis.org/mybatis-3/

[15] https://baike.baidu.com/item/junit/1211849?fr=aladdin

[16] https://docs.spring.io/spring/docs/5.1.0.BUILD-SNAPSHOT/spring-framework-reference

[17] http://jcp.org/en/jsr/detail?id=303

[18] https://blog.csdn.net/java_jsp_ssh/article/details/78483331

[19] https://blog.csdn.net/luanlouis/article/details/41408341